Risk Parity Fundamentals

Risk Parity Fundamentals

Edward E. Qian. PhD CFA

CRC Press is an imprint of the
Taylor & Francis Group, an **informa** business
A CHAPMAN & HALL BOOK

Front cover image: Binit Shah

CRC Press
Taylor & Francis Group
6000 Broken Sound Parkway NW, Suite 300
Boca Raton, FL 33487-2742

Printed on acid-free paper
Version Date: 20160105

International Standard Book Number-13: 978-1-4987-3879-8 (Hardback)

Visit the Taylor & Francis Web site at
http://www.taylorandfrancis.com

and the CRC Press Web site at
http://www.crcpress.com

Contents

Preface

IN SCIENCE, IT IS said that the process wherein a new and innovative idea gains acceptance has four stages*: (1) this is rubbish; (2) it is interesting but wrong; (3) it might only be relevant in certain rare cases; and (4) I told you so! This last phrase is not just a humorous punch line—it reflects the fact that an unconventional idea is not firmly established until its fieriest critics become converts and try to claim credits for it.

"Risk parity" was a new investment idea dating back to 2005. As an investment idea, even if quantitative in nature, it does not necessarily lie in the realm of science. Nevertheless, in its initial stages it has gone through similar processes before gaining acceptance from the investment community. The initial objection of risk parity stemmed from its use of leverage, which was regarded to be a sin after the 2008 global financial crisis. Later, the critics decried the allocation of bonds in risk parity portfolios based on tenuous views on the direction of interest rates. I would like to think that it has now reached or perhaps even passed the third stage where some investors are utilizing the risk parity approach in a limited way.† Would it ever reach the final stage where there is no more debate about its merits and all investors use risk parity on the entirety of their portfolios? It would never happen. By writing this book, however, I hope to convey the merit of risk parity in a broad sense. First, for some readers, the realization that risk allocation can play an active role in portfolio construction would be beneficial. For others, I hope to convince them to move past the third stage and consider the benefits of risk parity investing. Who knows? Some investors might even be emboldened enough to adopt risk parity for the entirety of their portfolios.

Risk parity, a term I christened in 2005, originated as a quantitative approach for asset allocation and it has since evolved in multiple directions. The first is the theoretical advance in the quantitative analysis of risk parity and risk-based investing in general. Many research papers, even a book, have been written on the technical aspects of risk parity portfolios. The second is practical applications of risk parity. Risk parity investing is not confined to asset allocation portfolios. For example, it is equally applicable to individual asset classes, multi-strategy portfolios, and multifactor models. This is attributable to the underlying principle of risk diversification in risk parity. Diversification has the ubiquitous ability to improve portfolio efficiency of many investment strategies.

* The original quote seems to belong to British biologist J. B. S. Haldane. I heard this version from Professor Alar Toomre at MIT.

† Behavior economics and finance have gone through similar stages.

These two evolutions are essential for the application of risk parity. The third aspect of evolution of risk parity, also relevant for its application and perhaps more crucial for its understanding and acceptance by the investment community, is the fundamental consideration of risk parity investing. Risk parity had its roots in directing investments based on quantitative analysis and modeling of investments risks. Like any quantitative technique, it does not function in a vacuum. Its core idea and applications raise many fundamental questions. For example, what are macroeconomic dimensions of risk in risk parity portfolios? What are the appropriate risk premiums in a risk parity portfolio? What are market environments in which risk parity might thrive or struggle? What is the role of leverage in a risk parity portfolio? These are just some of the questions facing investors. Many of our clients and prospects have asked us these questions before and after the launch of risk parity strategy by my team in 2006.

The answer to these questions helps put risk parity in the proper context in the world of investing. The reality is that unlike a scientific theory, a clear demonstration of risk diversification is often not sufficient to fully convince many investors of its practical merits. Investing is a multidisciplinary endeavor that combines theory and practice. It requires scientific and engineering inputs as well as historical and psychological considerations. Indeed, as we shall demonstrate, many questions regarding risk parity or other investment approaches are anecdotal and psychological in nature. Investors need a much broader perspective than just investment theory to address these questions and alleviate practical concerns. There is a fair chance that many will never be persuaded despite best efforts.

Over the last several years, with the help of my colleagues, I have written regular investment insights to both clarify and answer these questions mentioned above. Each article provides an insight or answer to address one specific question, using fundamental, quantitative, and historical analysis. No advanced degrees in quantitative fields are required for understanding these analyses.

This book is a collection of these investment insights. For example, one of the articles included is "Spear and Shield," (see Section 4.1 in Chapter 4) first written in July 2011, in a rebuttal to some critics of risk parity who object the inclusion of fixed income and commodities in a risk parity portfolio. One of the last research note included in this book was written in December 2014 on the relationship between forward and future interest rates. However, the articles are not arranged in chronological order. I have found it more effective to arrange them into chapters with common themes.

Chapter 1 covers the basic concept of risk parity and includes the original research article that coined the term risk parity. To facilitate a discussion of the fundamental aspects of risk parity, I also provide the basic elements of risk parity in terms of portfolio construction and characteristics. The rest of this book consists of investment insights regarding risk parity, grouped in seven chapters. Chapter 2 covers questions regarding risk premiums captured by risk parity portfolios. It is obvious that equity and interest rate risks should be included. But what about commodities and currencies? Do high-yield bonds belong to risk parity? Chapter 3 discusses the role of interest rate risk. There has been fierce criticism of risk parity because of its sizeable allocation to fixed income and many have proclaimed the "death" of interest rate risk premiums. We provide several perspectives on this issue.

Chapter 4 highlights the diversification benefits of risk parity portfolios. We provide several answers as to why risk parity portfolios could be truly diversified. When analyzing risk parity portfolios, it is important to see the forest for the trees. Chapter 5 deals with many practical details of risk parity portfolios, such as leverage, benchmark, portfolio rebalancing, and diversification return. We also highlight risk parity portfolios' ability to provide upside participation and downside protection. Chapter 6 reviews history lessons of risk parity in various market environments. What happened in 1994 when the Fed raised interest rates? What is the prospect of risk parity after the "taper tantrum" in the spring of 2013? Will the world follow the path of Japan? What happened to risk parity portfolios during 1970s? Risk parity is open to many different interpretations, some of which might be incorrect. Chapter 7 asks the question how investors can know whether an investment strategy is a risk parity strategy. We provide an answer with style analysis on several risk parity managers. The results indicate that some of them are not risk parity. Chapter 8 covers applications of risk parity. In addition to asset allocation, it can be applied to individual asset classes, as well as yield seeking fixed income portfolios. One could also consider risk parity as a global macro (GM) hedge fund. On a total portfolio level, how does it apply to both public and corporate pension funds in terms of asset/liability framework?

In the course of writing these investment insights, I have benefited enormously from discussions with my colleagues. Special thanks go to Bryan Belton, who not only coauthored some of the articles with me but also read all of them and provided valuable comments and suggestions. I have also had the pleasure to discuss risk parity investing with many investors regarding issues related to their challenges in meeting investment goals and benefited from their views on risk parity. I thank Michael Campbell and Bryan Hoffman for those discussions and their comments on many chapters in this book. John Linder and Hua Fan also provided valuable feedbacks. However, all errors are mine.

Author

Edward E. Qian, PhD, CFA, is the chief investment officer and head of research of Multi Asset Group at PanAgora Asset Management in Boston. Dr. Qian earned a BS in mathematics from Peking University and a PhD in applied mathematics from Florida State University. He was a postdoctoral researcher in astrophysics at the University of Leiden in the Netherlands and a National Science Foundation Postdoctoral Mathematical Research Fellow at MIT (Massachusetts Institute of Technology).

Dr. Qian started his investment career in 1996, first as a fixed-income quantitative analyst at Back Bay Advisors and then as a senior asset allocation analyst at Putnam Investments. Since 2005, Dr. Qian has been with PanAgora, where he is a member of the firm's senior management committee. Dr. Qian's investment research has been extensive and influential. His papers on financial interpretation of risk contribution and asset allocation laid the theoretical foundations for Risk Parity investment strategies. He coined the term "Risk Parity" and has published numerous papers on the topic of Risk Parity. Dr. Qian also made significant contribution to quantitative equity portfolio management. He pioneered in using portfolio theory for evaluating alpha factors and constructing multifactor models. He is the coauthor of the book *Quantitative Equity Portfolio Management: Modern Techniques and Applications* (Chapman & Hall, 2007). Dr. Qian is the recipient of Bernstein Fabozzi/Jacobs Levy awards for outstanding articles.

CHAPTER 1

An Introduction to Risk Parity Principle

THIS CHAPTER INTRODUCES RISK parity as a portfolio construction process. In this sense, risk parity is not merely a particular portfolio or strategy—it is a unified principle to direct all types of investments, ranging from asset allocation, to individual asset classes, to factor-based portfolios or models. The possibility is limitless. In order to address fundamental questions regarding risk parity portfolios and real-world applications of the risk parity principle, it is imperative to have a basic understanding of the tenets of risk parity, as a quantitative investment process. To this goal, we here address three topics: risk, risk contribution, and risk parity investing.

First, the risk. We note that concepts of risk and risk budgeting or risk contribution were not new when risk parity came on the scene. For example, risk management professionals had used these concepts and measures for quite some time. The big difference between risk management and risk parity investing is how one uses these measures to construct or influence portfolios.

Quantitatively, one can measure the risk of an investment by standard deviation or volatility of its returns when the returns can be adequately modeled by a normal distribution. For instance, broad market indices of developed country equities have annualized volatilities between 15% and 20% while market indices for high-quality fixed-income assets have much lower volatilities, close to 5%. Given sufficient data and/or structural models, one can estimate the risk of any investment in asset classes, country-specific investments, sectors, and individual securities. For returns that are far from normal, one might have to resort to risk measures that capture tail risk, such as value at risk (VaR).

The second concept is risk contribution, which is useful for risk analysis of portfolios. For portfolios of investments, whether asset allocation, equity, or fixed income, we can calculate or derive portfolio risks in terms of either volatility or other risk measures. For example, a typical 60/40 asset allocation portfolio with 60% in equities and 40% in fixed-income assets has annual volatility close to 10%. However, portfolio risk analysis does not stop here. For risk management and diversification consideration, one needs to know risk contributions from the underlying components. The question specific to the 60/40 portfolio would be how much of the 10% risk is attributable to equities and bonds, respectively. We shall present a simple calculation of this risk contribution exercise. The numbers will reveal a shockingly lopsided concentration in equity risk. However, what do risk contributions mean in economic terms? What does one do when confronted with such lopsided risk contribution?

The concept of risk contribution had also been used in many quantitative investment processes. For instance, in quantitative equity investing, portfolio managers actively monitor and budget risk contribution from systematic risks associated with risk factors such as value, size, and stock-specific risks. In fixed-income investing, portfolio managers pay attention to risk contribution from duration, term structure, credit, sector, and volatility. In multi-strategy hedge funds, one should care about risk contributions from different managers.

So if the concept of risk contribution was common practice in risk management, what is so special about risk parity? The innovation and defining characteristic of risk parity lies in its active use of risk contribution as the underlying criterion to construct portfolios, rather than as a limited and often passive tool for risk monitoring. Using an asset allocation portfolio with equities and fixed income as an example, risk parity, in a strict sense of parity, devoid of any active views of the market, seeks a portfolio that has 50% of its risk from equities and 50% from fixed income. Once a portfolio, with a specific total portfolio risk target, is found, that is the portfolio to invest—no questions asked.

While this evolution in portfolio construction through risk allocation seems innocuous, it has some significant implications and consequences. We shall mention some here because they are often sources of questions from market participants regarding applications of risk parity. First, at a glance, risk parity seems to do away with return forecasting since only risk inputs are required. This is not entirely true. On a long-term basis, if risk parity portfolios are assumed optimal, they imply approximately equal risk-adjusted returns for underlying portfolio components. Second, on a tactical basis, one can still utilize return forecasts to adjust risk contribution, away from parity, to incorporate active views. Nevertheless, the influence of return forecasting, which can be highly unreliable judging from the performance of many active managers, has been reduced.

The second consequence of risk parity is its exposure to low-risk asset classes, sectors, and securities. Equal risk allocation requires the notional exposure to low-risk assets to be higher relative to high-risk assets. Interest-rate exposure from fixed-income assets in asset allocation portfolios is one example but it is not the only example by any means. How would one think about such exposures and the related issue of portfolio leverage?

Third, risk parity is agnostic to conventional benchmarks and capitalization-weighted indices. While one of the premises of risk parity investing is that superior diversification benefits allows it to outperform these traditional capital-based investments over the long run, risk parity portfolios in general do not and probably should not track conventional indices. This brings out two issues. One is what should be the appropriate short- and long-term benchmarks for risk parity portfolios. The other is, are risk parity portfolios passive or active?

We shall address these and other questions in the remaining chapters of this book. For the remainder of the chapter we introduce the definition of risk contribution and provide a financial interpretation of this quantitative concept. We base our discussion on the article "On the Financial Interpretation of Risk Contribution: Risk Budgets Do Add Up" (Qian, 2006), which establishes the relationship between risk contribution and return contribution, thus putting risk parity on a sound theoretical foundation. We then use a simple example to illustrate the risk parity portfolio construction process and highlight the difference between naïve risk parity and risk parity. We then present the investment insight that applies risk parity to asset allocation portfolios, coining the term "risk parity."

1.1 RISK CONTRIBUTION AND ITS FINANCIAL INTERPRETATION

1.1.1 Risk Contribution

Let us start with the old way of going from portfolio weights to portfolio risk and risk contribution. Portfolios with two assets are the easiest to analyze. Suppose we have a 60/40 portfolio with 60% in stocks and 40% in bonds. We further assume equity volatility is 15%, fixed-income volatility is 5%, and the correlation between the two assets is 0.2. Based on these assumptions, the total risk of the portfolio, in terms of volatility, is

$$\sigma = \sqrt{(0.6)^2(0.15)^2 + 2 \cdot 0.2 \cdot (0.6) \cdot (0.4) \cdot (0.15) \cdot (0.05) + (0.4)^2(0.05)^2} \approx 9.60\%. \quad (1.1)$$

Under the squared-root bracket, the first and the third terms are variances from stocks and bonds respectively, and the middle term is the covariance. It is equivalent to state the risk in terms of variance, which is volatility squared. In this case, we have

$$\sigma^2 = (0.6)^2(0.15)^2 + 2 \cdot 0.2 \cdot (0.6) \cdot (0.4) \cdot (0.15) \cdot (0.05) + (0.4)^2(0.05)^2 \approx (9.60\%)^2. \quad (1.2)$$

How do we define risk contribution from stocks and bonds respectively, using the variance in Equation 1.2? It is easy to see that the first term $(0.6)^2(0.15)^2$ is attributable to stocks and the last term $(0.4)^2(0.05)^2$ belongs to bonds. What about the covariance term? It turns out that splitting half-half is the correct way. Proceeding accordingly, the contribution to variance from stocks is

$$\sigma_s^2 = (0.6)^2(0.15)^2 + 0.2 \cdot (0.6) \cdot (0.4) \cdot (0.15) \cdot (0.05). \quad (1.3)$$

The contribution from bonds is

$$\sigma_b^2 = (0.4)^2(0.05)^2 + 0.2 \cdot (0.6) \cdot (0.4) \cdot (0.15) \cdot (0.05). \quad (1.4)$$

We can see the sum of Equations 1.3 and 1.4 equals the total in Equation 1.2.

With this variance decomposition, we can now calculate the percentage of risk contribution from stocks and bonds, respectively. For stocks, we have

$$p_s = \frac{\sigma_s^2}{\sigma_s^2 + \sigma_b^2} \approx \frac{(0.6)^2(0.15)^2 + 0.2 \cdot (0.6) \cdot (0.4) \cdot (0.15) \cdot (0.05)}{(9.60\%)^2} = 92\%. \tag{1.5}$$

For bonds, we have

$$p_b = \frac{\sigma_b^2}{\sigma_s^2 + \sigma_b^2} \approx \frac{(0.4)^2(0.05)^2 + 0.2 \cdot (0.6) \cdot (0.4) \cdot (0.15) \cdot (0.05)}{(9.60\%)^2} = 8\%. \tag{1.6}$$

Hence, based on the assumed risk inputs, the risk contribution from stocks and bonds are 92% and 8%, respectively. In other words, the majority of risk in a 60/40 portfolio is due to equity. This is not exactly what a so-called "balanced" portfolio is supposed to be.

We can look at the dominance of equity risk in a 60/40 portfolio from another perspective. Suppose a 60/40 portfolio invests 60% in stocks, and 40% in cash instead of in bonds. With cash in the portfolio, we can view it as de-levered equity portfolio, from 100% stocks and 0% cash to 60% stocks and 40% cash. There is no diversification to speak of in this stock/cash portfolio. What is the risk of the 60/40 stock/cash portfolio? It is

$$\sigma = (0.6)(0.15) = 9.0\%. \tag{1.7}$$

This is very close to the 9.6% risk of the 60/40 stock/bond portfolio. In other words, the volatility of bonds is so low compared to stocks that its impact on the portfolio is almost as small as cash, which has zero volatility.

1.1.2 A Three-Asset Example

Risk contribution for portfolios with more than two assets can be worked out in a similar fashion. We give one more example with three assets with the additional asset serving as a proxy for commodities. This three-asset example will be revisited again in Chapter 5. Commodities have higher volatility than stocks and bonds. Its correlations with them vary with inflation. In a low inflation regime, commodities tend to have negative correlation with nominal bonds and positive correlation with stocks. In a high and/or rising inflation regime, commodities tend to have negative correlations with both stocks and bonds. For the purpose of the example, we assume the return volatility of commodities is 25% and its correlations with stocks and bonds are 0.2 and −0.2, respectively. Table 1.1 summarizes the risk estimates.

Let us consider a 40/40/20 portfolio, with 40% in stocks, 40% in bonds, and 20% in commodities. The portfolio risk is approximately 8.82% and the variance is 8.82% squared. To derive risk contribution of one particular asset, based on the variance, we would sum up

TABLE 1.1 Volatility and Correlation Assumptions for Three-Asset Classes

	Bonds	Stocks	Commodities
Volatility	5%	15%	25%
Bonds	1.0	0.2	–0.2
Stocks	0.2	1.0	0.2
Commodities	–0.2	0.2	1.0

variance of that asset and halves of its covariances with the other two assets. For stocks, we have

$$\sigma_s^2 = (0.4)^2(0.15)^2 + 0.2 \cdot (0.4) \cdot (0.4) \cdot (0.15) \cdot (0.05) + 0.2 \cdot (0.4) \cdot (0.2) \cdot (0.15) \cdot (0.25). \quad (1.8)$$

The first term is the variance due to stocks on its own, the second term is the half of the covariance between stocks and bonds, and the last term is the half of the covariance between stocks and commodities. Similarly, we can calculate contributions from bonds and commodities, σ_b^2 and σ_c^2.

The percentage risk contributions are then given by the ratio of variance contribution to the total variance. For instance, the risk contribution of stocks is

$$p_s = \frac{\sigma_s^2}{\sigma_s^2 + \sigma_b^2 + \sigma_c^2} \approx 57\%. \quad (1.9)$$

We can also calculate the remaining risk contributions from commodities and bonds. From commodities, it is $p_c \approx 37\%$ and from bonds, it is $p_b \approx 6\%$. We note that compared to the original 60/40 portfolio redirecting 20% from stocks to commodities reduces the risk contribution from stocks to 57% but does little to change bonds' low-risk contribution.

These risk contribution results all show that the weights of asset allocation portfolios consisting of assets with vastly different risk levels often do not resemble risk contributions to portfolio risk. They reveal that a so-called balanced portfolio is not at all diversified and investing in "alternatives" such as commodities does little to improve risk balance between growth assets (stocks and commodities) and safe assets (high-quality bonds). There is no balance in risk in a so-called balanced portfolio.

Wait a minute, someone might say. These risk contribution calculations are mathematically derived. How do we know that they actually mean anything related to portfolio risk and portfolio diversification?

1.1.3 Financial Interpretation of Risk Contribution

In fact, as recent as about 11 years ago, questions and doubts regarding the validity of risk contribution were rampant among financial economists and investment experts. The basic issue was many were familiar with marginal contribution to risk but they could not quite grasp the concept of risk contribution.

Economists love marginal "things"—no pun intended. For example, marginal utility is the change in utility when the consumption of goods or investors' wealth changes. There is marginal product cost of one additional unit, etc. Mathematically, one assumes a functional form of utility, then the marginal utility is expressed as a derivative—relative change of the dependent variable versus the independent variables, of utility with respect to wealth, for example. When it comes to portfolio risk, marginal contribution to risk measures the change in portfolio risk when the weight of an asset in the portfolio changes from the current weight. This concept, in turn, can also be expressed as a partial derivative of portfolio risk with respect to weights. For instance, we might be interested in how the risk of a 60/40 portfolio would change when we tweak the weight of stocks or bonds ever so slightly.

In a sense, marginal contribution to risk is a dynamic concept. On the one hand, risk contribution is about contribution to the current state; or borrowing another term from classical mechanics, it is a kinematic concept. This difference caused many to question and criticize its use in investment management. For instance, Sharpe (2002) argues that a mere mathematical decomposition of risk does not necessarily qualify as risk contribution and went on to suggest rejecting the concept of risk contribution altogether. Chow and Kritzman (2001) express a similar critical view toward risk budgeting while emphasizing the usefulness of marginal contribution to VaR because of its clear financial interpretation. Others (Grinold and Kahn, 2000; Litterman, 1996) had acknowledged the usefulness of risk contribution but struggled to interpret it using marginal contribution to risk.

Contrary to these criticisms and misinterpretations, risk contribution does have independent, intuitive financial interpretations. Risk contributions to portfolio risk are directly linked to return contributions to portfolio returns from underlying assets.

This linkage is twofold. One is on average return and the other is on tail risk. First, average return contributions from underlying assets over the long run are equal to risk contributions, when average returns are commensurate with risks or when the portfolio is mean-variance optimal. When this is the case, a procedure called reverse optimization can be used to derive equilibrium returns for the assets in the portfolio.

This is better illustrated with an example. Let us consider the 60/40 stock/bond portfolio. Were the 60/40 portfolio optimal, the expected returns of stocks and bonds must follow certain pattern, determined by a covariance matrix and their portfolio weights. In the 60/40 example, suppose we fixed the expected return of bonds at 2%, then the expected return of stocks must be 14.8%, given the fact that the portfolio has 92% of its risk in stocks. Leaving aside the seeming improbability of this extreme level of equity risk premium, we can calculate return contributions from stocks and bonds. The portfolio expected return equals

$$r_p = (0.6) \cdot (14.8\%) + (0.4) \cdot (2.0\%) = 8.88\% + 0.80\% = 9.68\%. \tag{1.10}$$

Thus, the contribution from stocks is 8.88%/9.68% = 92% and the contribution from bonds is 0.80%/9.68% = 8%, which exactly match the risk contributions of stocks and bonds.

This linkage between risk contribution and return contribution has important implications regarding the optimality of the 60/40 portfolio. First, if one assumes the 60/40 portfolio is optimal then one is expecting 92% of its return to come from stocks and does not expect bonds to contribute much at all. This implies that there is little diversification of return contribution. If we hold stocks to capture equity risk premium and bonds to capture interest rate risk premium, then the 60/40 portfolio is mostly capturing equity risk premium.

Second, the implied equity risk premium over bonds is much too high compared to its historical average, which is around 4%. When the expected future returns are more in line with the historical average, then the 60/40 portfolio cannot possibly be an optimal asset allocation portfolio. Simple logic would dictate that to make the portfolio more optimal, one must increase the allocation to bonds and decrease the allocation to stocks. However, we are getting ahead of ourselves.

1.1.4 Risk Contribution and Loss Contribution

The second linkage is more about risk. The risk contribution is a close approximation of loss contribution when the portfolio suffers a significant loss. One of the common pressing questions facing investors is: In the event of a sizable loss to a portfolio, what are the likely contributions to the loss from the underlying components of the portfolio? The answer to this question turns out to be consistent with the risk contribution we have defined.

Let us get more specific on this point. Suppose a portfolio had a substantial loss in terms of a large negative return, called L. We are asked to tell how much of the loss is attributable to each of the underlying asset classes. In the case of the 60/40 portfolio, the question is what percentage of loss is from stocks and bonds, respectively. We might write loss contributions as L_s/L for stocks and L_b/L for bonds. Then, it can be proved mathematically (Qian 2006) that

$$L_s/L \approx p_s, \quad L_b/L \approx p_b. \tag{1.11}$$

This is true for both standard deviation of return and VaR as the risk measure.

For technical detail, we refer interested readers to Qian (2006). We provide some intuition here as to why this relationship is true. First, risk contributions are similar to betas of individual assets versus the overall portfolio. Second, the loss contribution, based on conditional distribution, is a function of both beta and average return. Third, when the loss is significant, the average returns become negligible in comparison. Therefore, we are left with beta or risk contribution only. Fourth, the accuracy of this relationship should improve as the loss becomes bigger.

This relationship between risk contribution and loss contribution validates the concept of risk contribution. It is the theoretical foundation of risk parity portfolios, at least from a risk management perspective. In fact, it is this very idea that prompted the author to create the notion of risk parity. Of course, risk parity can also be thought of in other contexts, which would be the subjects of later chapters. I must also add that on the other side, risk contribution is also related to profit contribution when the profit is large enough.

TABLE 1.2 Monthly Return Statistics of Indices and the 60/40 Portfolio

	S&P 500	US LT Government	60/40 Portfolio
Average return	0.98%	0.46%	0.78%
Standard deviation	5.61%	2.27%	3.61%
Correlation to S&P 500	1.00	0.14	0.97

For example, in a 60/40 portfolio, one should expect that large portfolio gains are mostly due to stocks returns.

How accurate is this linkage between risk contribution and loss contribution? We compare theoretical predictions with the actual results of a 60/40 portfolio investing 60% in the S&P 500 index and 40% in Ibbotson's long-term (LT) government bond index. The monthly returns span from January 1926 to June 2004.* Table 1.2 gives the statistics of monthly return indices, as well as returns of the 60/40 portfolio. The reason for using monthly statistics rather than annual statistics is we shall calculate contribution to losses in calendar months. The monthly average returns of stocks and bonds are 0.98% and 0.46%, respectively, the monthly standard deviations are 5.61% and 2.27%, and their correlation is 0.14. The balanced portfolio has an average return of 0.78% and a standard deviation of 3.61%. By definition of risk contribution, the percentage contributions to risk are 90.3% from stocks and only 9.7% from bonds, similar to our previous hypothetical example.

With monthly returns of the two underlying assets and the portfolio, we can now compute contribution to losses. Since the monthly returns of the portfolio are discrete, we group them into several bins, starting with losses exceeding –3%. The bins are listed in the first column of Table 1.3. For each bin, we select all the returns of the 60/40 portfolio that fall in the bin. For example, there were 45 months when the 60/40 portfolio delivered returns in the bin from –3% to –4%. For each month, we then calculate contributions from stocks and bonds to the particular loss in the month. Finally, we take the average of the loss contributions over the 45 months. Table 1.3 shows the percentage contributions to loss with the realized values for the balanced portfolio. In each bin, we only report contribution from stocks.†

TABLE 1.3 Realized Loss Contributions from Stocks of the 60/40 Stock/Bond Portfolio

Loss (%)	Loss Contribution (%)	Number of Months
–4 to –3	89.8	45
–5 to –4	92.7	23
–6 to –5	88.1	11
–7 to –6	99.5	9
–8 to –7	90.1	8
Below –8	102.4	12

* Originally written by the author in 2005.

† Since the two percentage contributions add up to 100%, the bond contributions are omitted.

Table 1.3 shows that the realized loss contribution from stocks is in general very close to the theoretical value of 90.3% except for two bins. The error is especially large in the last bin with losses exceeding −8%. The realized value is above 102%. We provide two possible reasons for the poor accuracy at the tail end of the return distribution. One, there is not enough data points, which might also explain the mismatch in the loss bin from −8% to −7%. The other is that stock returns have fat tails. The risk contribution based on variance underestimates contribution from stocks in the left tail.*

We also note that when the loss exceeds −8% the contribution from stocks is above 100%. This implies, of course, that bonds' contribution is negative in those circumstances, which correspond to severe financial stress in capital markets. Yet, bonds usually benefit from flight to safety in those situations and deliver positive returns on average (hence the negative contribution to losses that are negative). However, because bond's risk contribution is so low to begin with, its positive return barely made a dent in the huge losses incurred by stocks.

In general, if an asset has more left tail risk associated with negative skewness and excess kurtosis, then its loss contribution would be higher than what risk contribution indicates with variance being the risk measure. Within asset allocation portfolios, these assets are stocks (more so for emerging market stocks), high-yield (HY) bonds, and emerging market bonds. While this issue deserves some attention, however, the problem is much more severe for many hedge fund strategies. In my view, using variance as the risk measure is justifiable for asset allocation portfolios if one chooses asset classes judiciously and improves the distribution of underlying asset classes with diversification. But it is not justifiable to use variance as the risk measure for many hedge fund strategies due to their debilitating tail risks.

To summarize, both theoretical proof and empirical evidence show that risk contribution has a sound economic interpretation—either as average return contribution for a mean-variance optimal portfolio or as expected contribution to potential large losses of any portfolio. When the underlying return distributions are normal, risk contribution in terms of variance is easy to calculate and often adequately depicts the loss contribution. Risk contribution in terms of VaR, on the other hand, is precise in theory, but hard to compute in practice. It is most useful for certain hedge fund strategies. All subsequent analyses and examples in this book will be based on risk contribution with variance.

1.2 RISK PARITY

Once the relationship between risk contribution and return and loss contribution is established, the potential benefits of using risk contribution as a tool to construct portfolios become rather apparent. Instead of computing risk contribution of a given portfolio and perhaps lamenting over its risk concentration, why not start with a balanced risk allocation and seek the portfolio that matches the prescribed risk allocation? This is the essence of risk parity. We now describe how this is done in principle.

* I studied the estimation of contribution to VaR for the same 60/40 portfolio and found that, indeed the prediction became more accurate when the losses exceed −6% (Qian, 2006).

1.2.1 Risk Parity Portfolio Construction

Let us start from a typical mean-variance optimization process, in which the inputs are both forecasted returns and risk estimates of volatilities and correlations. The mathematical process is to maximize a utility function based on the expected return and risk of the portfolio. While this is a well-defined mathematical problem, its outcome is often a highly concentrated portfolio. This is because the optimization procedure is highly sensitive to both return and risk inputs. Any minute change in both can cause seismic changes in the optimized portfolio weights. These drawbacks often cause mean-variance portfolios to be unintuitive and hence hard to use in practice.

Risk parity portfolio construction is different in two key aspects. First, it does not rely on return forecast per se. Without any active views on asset returns or more precisely risk-adjusted returns, equal risk allocation or parity becomes the starting point of portfolio construction. However, one can incorporate active views by adjusting the risk allocation away from parity. We would still refer to these portfolios as risk parity portfolios in a general sense as long as the adjustments do not result in a portfolio that is not risk balanced.

The second aspect of risk parity is it does not use an optimization process. As a result, risk parity portfolio weights are less sensitive to risk inputs. The portfolio weights are quite intuitive in the sense that they reflect risk inputs—the higher (lower) the volatility the lower (higher) the weight and the higher (lower) the correlation the lower (higher) the weight.

Let us illustrate these points with a few examples. First, for two-asset portfolios, the weights of risk parity portfolios are inversely proportional to volatility, regardless of their correlation. If stocks and bonds have volatility of 15% and 5%, respectively, with a ratio of 3:1, then the ratio of portfolio weights is the reciprocal 1:3. In other words, for each 1% weight in stocks we must have 3% weight in bonds to balance the risk contribution. For a fully invested portfolio, it works out to have 25% in stocks and 75% in bonds. This is a rather unremarkable portfolio, or as it is often referred to, a conservative 25/75 portfolio. Its returns are lower but more stable than that of 60/40 portfolios. In other words, it has lower expected return but a better Sharpe ratio. Investors prefer a good Sharpe but still need good returns just as people prefer a good diet but need enough calories to live. To get that return, a risk parity portfolio would lever this 25/75 portfolio such that the risk is comparable to that of 60/40. The result is a 50/150 portfolio with a leverage of 200%, financed by a short position of 100% in cash.

Figure 1.1 gives a schematic illustration of the two approaches. The line for the traditional asset allocation portfolios flattens out to right of Figure 1.1 as the stock risk dominates and diversification benefits disappear. In contrast, the line for risk parity portfolios is straight maintaining the linear relationship between risk and excess return since the portfolios always maintain the same risk diversification regardless of the level of portfolio risk.

1.2.2 Risk Parity Portfolios with Three Assets

For portfolios with more than two assets, one could still opt to determine portfolio weights by the inverse of volatility without considering correlations. The result is a portfolio that is not truly risk parity. It only leads to risk parity if all pairwise correlations are the same. We shall refer to them as naïve risk parity portfolios.

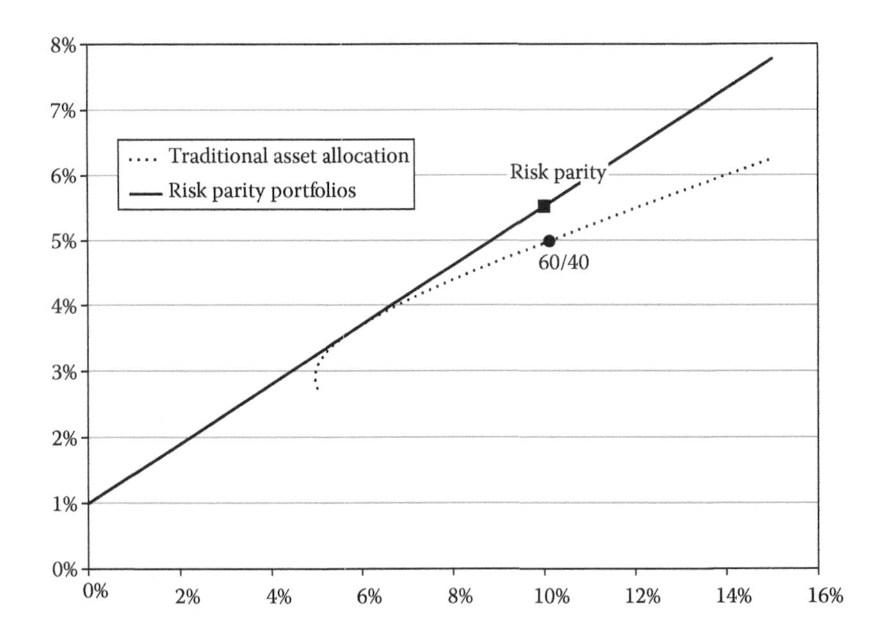

FIGURE 1.1 Traditional asset allocation portfolios versus risk parity portfolios.

For example, consider risk parity portfolios for the three-asset classes whose risk metrics are listed in Table 1.1. With naïve risk parity, the portfolio weights would be 65% in bonds, 22% in stocks, and 13% in commodities, as shown in Table 1.4. The weights sum up to 100% so the portfolio is not levered. They are in the reverse order of asset volatilities.

However, this portfolio is not truly risk parity. Our calculation of risk contribution shows that stocks have more risk contribution than the other two assets. This is because stocks have positive correlations with both bonds and commodities. In order to be risk parity, the allocation to stocks should be lower. It is easy to find the weights of a true risk parity portfolio by trial and error when dealing with three assets. One can also use Microsoft excel solver for the job when the number of assets is not too many. In this case, the weights turn out to be 68% in bonds, 18% in stocks, and 14% in commodities.

We make two remarks about the differences between two sets of portfolio weights. First, the changes make intuitive sense. The stocks weight is lower because of its relatively high correlations with other assets. Second, at first glance these changes might appear insignificant. This impression is incorrect for a couple of reasons. One, the risk parity portfolios are often levered 2:1. When this is done, the difference between the weights of stocks is 8%, which can make quite an impact on the portfolio returns. Two, the correlations listed in Table 1.1 are quite small in magnitude. When the magnitudes of correlations are larger,

TABLE 1.4 Portfolio Weights for Naïve Risk Parity and Risk Parity Portfolios

	Bonds (%)	Stocks (%)	Commodities (%)
Naïve risk parity	65	22	13
Risk parity	68	18	14

their effect on portfolio weights would be stronger. For instance, in the aftermath of the 2008 global financial crisis, the capital markets experienced many risk on/risk off episodes, whereas the correlation between stocks and commodities became extremely high and the correlations between high-quality bonds and the two risky assets became quite negative. A true risk parity portfolio would have more exposure to high-quality bonds than a naïve risk parity portfolio does. This would have benefited the true risk parity portfolio tremendously since high-quality bonds have delivered the best Sharpe ratio among the three assets in the aftermath of the global financial crisis.

The following research note reiterates some of the analysis we presented above and then applies it to stock/bond asset allocation portfolios. It illustrates why a leveraged risk parity portfolio with a targeted risk can be expected to outperform the 60/40 portfolio.

1.3 EFFICIENT PORTFOLIOS THROUGH TRUE DIVERSIFICATION*

1.3.1 Eggs in One Basket

One well-understood and seemingly well-heeded axiom on investing is "Do not put all your eggs in one basket." So, if you were advised to place over 90% of your eggs in one basket, would you think that is sufficient diversification? Apparently, many investors, those who invest in a balanced portfolio of 60% stocks and 40% bonds, do.

How can this be true? The answer is size matters—the stock "eggs" are about *nine* times as big as the bond "eggs," when considering risks of each asset class. Assume stock and bond returns have a standard deviation of 15% and 5%, respectively, on an annual basis. Then, in terms of variance, which is the correct measure to use in analyzing portfolio risks, stocks are nine times as risky as bonds. For now, just imagine we have six stock "eggs" of size nine and four bond "eggs" of size one in two separate baskets. In total, we have an equivalent of $6 \cdot 9 + 4 = 58$ "eggs," of which 54 are from stocks; 54 out of 58 is about 93%.

While our egg analogy might appear simplistic, it is not far from reality.† For example, from 1983 to 2004 the excess return of the Russell 1000 index had an annualized volatility of 15.1% and the excess return of the Lehman‡ Aggregated Bond index had an annualized volatility of 4.6%, while the correlation between the two was 0.2. Based on these inputs, our calculation shows that for a 60/40 portfolio, stocks contributed 93% of the risk and bonds contributed the remaining 7%. The message is clear—while a 60/40 portfolio might appear balanced in terms of capital allocation, it is highly concentrated from the perspective of risk allocation.

1.3.2 From Risk Contribution to Loss Contribution

Why should investors care about risk contribution? As we have shown in the previous section, risk contribution is a very accurate indicator of loss contribution. Going back to our

* Originally written by the author in September 2005.

† While the egg analogy is common in the investment industry, it is not common to see eggs that are nine times bigger than an ordinary egg. A more apt analogy suggested by my colleague Bryan Belton (pers. comm.) is that stocks and bonds are like hard liquor and beer, respectively. For someone who drinks six cups of liquor and four cups of beer using cups of the same size, it is obvious that contributions to the loss of sobriety are not balanced.

‡ The Lehman indices are now Barclays indices.

TABLE 1.5 Average Loss Contribution for the 60/40 Portfolio Based on Russell 1000 and Lehman Aggregate Bond Index: 1983–2004

Loss Greater Than (%)	Stocks (%)	Bonds (%)	*N*
2	95.6	4.4	44
3	100.1	−0.1	25
4	101.9	−1.9	14

previous example of a 60/40 portfolio, Table 1.5 displays the average contribution to losses with three different thresholds.

For losses above 2%, stocks, on average, contributed 96% of the losses. This is very close to the risk contribution of 93% we calculated earlier. For losses greater than 3% or 4%, the contributions from stocks are above 100%. While influenced by sampling error (N is the number of monthly returns for a given threshold) and higher tail risks from stocks, Table 1.5 provides empirical evidence for the economic interpretation of risk contribution—it is approximately the expected loss contribution from underlying components of the portfolio. This is true when one uses variances and covariances to calculate risk contribution.

1.3.3 Risk Parity Portfolios

It can now be understood why a 60/40 portfolio is not a well-diversified portfolio. Another way to look at Table 1.5 is the diversification effect of bonds is insignificant in a 60/40 portfolio. When a loss of decent size occurs, over 90% is attributable to stocks. Conversely, this would imply that any large loss in stocks leads to a large loss for the whole portfolio. This is hardly diversification.

How can we use these insights to design a portfolio that limits the impact of large losses from individual components? This can be accomplished if we make sure that the expected loss contribution is the same for all components. The concept of risk contribution and its economic interpretation thus lead us to the development of "risk parity" portfolios, a discipline that allocates risk equally among asset classes.

While "risk parity" portfolios can utilize many asset classes,* it helps to illustrate its benefit using the stock/bond example. For a fully invested portfolio, an allocation of 23% in Russell 1000 index and 77% in Lehman aggregate index would have equal risk contribution (ERC) from stocks and bonds. Table 1.6 reports some of its return characteristics along with those for the underlying indices as well as the 60/40 portfolio, all measured by

TABLE 1.6 Return Characteristics of Indices and Portfolios: 1983–2004

	Russell 1000	Lehman Agg.	60/40	Risk Parity I
Average return	8.3%	3.7%	6.4%	8.4%
Volatility	15.1%	4.6%	9.6%	9.6%
Sharpe ratio	0.55	0.80	0.67	0.87

* They should include asset classes that capture risk premiums from equity risk, interest rate risk, and inflation risk. Within each risk premium there could be further diversification by including subasset classes.

TABLE 1.7 Average Loss Contribution of the Unlevered Risk Parity Portfolio

Loss Greater Than (%)	Stocks (%)	Bonds (%)	*N*
2	48.4	51.6	17
3	45.4	54.6	6

excess return over 3-month treasury bills (T-bills). As we can see, the Russell 1000 index had the highest average return at 8.3% but with a much higher standard deviation, and as a result, it had the lowest return-risk or Sharpe ratio, at 0.55. The bond index had a lower average return and lower standard deviation but the Sharpe ratio, at 0.80, was better. For the 60/40 portfolio, both the average return and the standard deviation were between those of stocks and bonds. More importantly, its Sharpe ratio, at 0.67, was lower than that of bonds, another indication of poor diversification: the overall portfolio's Sharpe ratio was lower than one of its components.

In contrast, the Sharpe ratio of risk parity portfolio was higher than that of stocks and bonds, representing the benefit of true diversification. Note that it was substantially higher than that of the 60/40 portfolio.* With the same level of risk achieved with a modest degree of leverage, it outperformed the 60/40 portfolio by an average of 2% per year.

What does happen to loss contribution of the risk parity portfolio? Of course, we should expect that they are roughly equal for stocks and bonds. Table 1.7 shows for loss at 2% or above, stocks contributed 48% while bonds 52%; for loss at 3% or above, stocks contributed 45% while bonds 55%. These numbers are close to parity, given the limited numbers of sample points.

1.3.4 "Optimality" of Risk Parity Portfolios

In contrast to a traditional asset allocation approach, which typically involves forecasting long-term asset returns, employing mean-variance optimization, risk parity portfolios are purely based on risk diversification. The question is then why should risk diversification, especially balanced risk contribution, lead to efficient portfolios.

It turns out that risk parity portfolios are actually mean-variance optimal under certain assumptions. For two-asset portfolios such as stock/bond portfolios, the requirement is that both assets have the same Sharpe ratio.† Are these realistic assumptions? First, equal Sharpe ratio implies that the expected return is proportional to the risk for each asset class. This is theoretically appealing when assets are priced by their risk. The historical return data from the United States shows that the Sharpe ratios of large cap stocks and intermediate government bonds are roughly equal. Both theoretical argument and empirical evidence strongly suggests that risk parity portfolios over the long run might be optimal or very close to be optimal.

---

* One way to interpret the Sharpe ratio is the return in percentage point for every 1% of risk taken. For example, for every 1% risk taken, the 60/40 portfolio returns 0.67% while the parity portfolio returns 0.87% per annum.

† For portfolios with more than two assets, one condition is all assets have the same Sharpe ratio and all pairwise correlations are identical.

In practice, we can derive the implied return from an equal Sharpe ratio assumption and note the results are quite realistic. For instance, if we assume a Sharpe ratio of 0.3, then the implied excess returns (over the risk-free rate) would be 4.5% and 1.5%, respectively, for stocks and bonds. An equity risk premium of 3% over bonds is slightly lower than its historical average. These considerations lead us to believe that the parity portfolios are efficient, not only in terms of allocating risk, but also in the classical mean-variance sense under the given assumption.

Risk parity portfolios have additional benefits. First, each asset is guaranteed to have nonzero weight in the portfolios. In contrast, portfolios derived with mean-variance optimization could be extremely concentrated. Second, the weights are influenced by asset return correlations in a desirable way: assets with higher (lower) correlations with others will have lower (higher) weight, other things being equal. This could affect the weight of commodities in the portfolios since its correlations with stocks and bonds could be quite low.

1.3.5 Targeting Risk/Return Level with Appropriate Leverage

While the Sharpe ratio of the risk parity portfolio is high, a fully invested version would have low risk and insufficient return. One cannot simply live on the Sharpe ratio. To achieve higher levels of return, an appropriate amount of leverage is required. The risk parity portfolio in Table 1.6 has a leverage ratio of 180%. Table 1.8 shows one additional version of leveraged risk parity portfolio, whose risk level is the same as that of stocks. It outperformed the Russell 1000 by close to 5% per year with a leverage ratio of 280%.

Leverage is necessary and relatively easy to implement with exchange-traded futures. When the targeted risk is not too high, it is appropriate as well as prudent. We shall have more discussion about the issue of portfolio leverage in Chapter 5. One popular but misguided perception is one is only leveraging the low-risk bonds while leaving stocks alone.* A total portfolio perspective is more accurate: we have built the risk parity portfolio with a high Sharpe ratio and we apply leverage on the whole portfolio rather than on any single component.

1.3.6 Using Risk Parity Portfolios

Risk parity portfolios can be used as stand-alone beta portfolios and they can be combined with alpha strategies to further increase returns. As a beta portfolio, a risk parity portfolio

TABLE 1.8 Comparison between Parity Portfolios with the Stock Index and the 60/40 Portfolio

	Russell 1000	Risk Parity II	60/40	Risk Parity I
Average return	8.3%	13.2%	6.4%	8.4%
Volatility	15.1%	15.1%	9.6%	9.6%
Sharpe ratio	0.55	0.87	0.67	0.87%

* Stocks have embedded leverage due to corporate debts.

can be used at least in three ways. The first is an unleveraged version with 4%–5% risk, similar to that of Lehman Aggregate Bond Index. The second is a leveraged version with leverage ratio about 2:1 and a risk target around 10%, similar to that of domestic or global balanced portfolios. The third approach is to use it as a global macro (GM) strategy with 20% risk and leverage of 4:1. Our back tests show the risk parity portfolio had a Sharpe ratio of 1.1 over the period from 1983 to 2004, translating to excess returns of 4.5%, 11.3%, and 22.6%, respectively, for the three versions.

What would the future bring? Given the current challenges of a low return environment, investors must seek reliable alpha sources as well as superior beta returns. For many investors, the beta risk actually represents the majority of their total risk budgets. Risk parity portfolios provide a more efficient alternative to traditional asset allocation. It limits the risk of overexposure to any individual asset while providing ample exposures to all assets. With a risk parity portfolio, we can reap the benefit of true diversification. In a way, the eggs are now placed evenly and safely in many baskets.

CHAPTER 2

The "Colors" of Risk Premiums

THE INVESTMENT WORLD IS full of investment products that compete for investors' attention and money. For instance, there are more equity mutual funds than there are individual stocks. Meanwhile, Wall Street financiers are constantly searching for the next new thing that might meet investors' needs or catch their fancies. Like a child in a candy store or a homemaker deciding the color of carpet, it could be quite confusing for investors to decide where to invest their money.

Things appear not so dire for asset allocation investors. It used to be just stocks and bonds. Now it is stocks, bonds, and alternatives, to be a little simplistic. However, as we dig deeper, the number of sub-asset classes quickly adds up. In stocks, there are stocks with different styles (large/mid/small and growth vs. value), and different geographic region (US/non-US, developed, and emerging). Bonds are more numerous: bonds with different credit quality (investment grade [IG] to high yield [HY]), different maturity (short/intermediate/long term), and different geographic region (developed to emerging). The alternative bucket is even more diverse: commodities, hedge funds, private equity, real estate, and infrastructure. In other words, anything goes. One of the common questions from investors is, what asset classes should be included in a risk parity portfolio?

In this regard, it may help to think in colors. The natural world, to a human eye, is filled with wondrous and a seemingly limitless spectrum of colors. However, it turns out most of what we see, due to physiological reasons, is based on combinations of just "three" primary colors: red, green, and blue—the RGB model of colors.

By way of analogy, there are "three" primary risk premiums in the space of liquid asset classes. They are equity risk premium, interest rate risk premium, and inflation risk "premium." These are the primary "colors" of risk premiums. A risk parity portfolio should at least have balanced exposures to these three risk premiums.

There are two reasons to consider asset allocation investing along this line. First, these three risk premiums are the major sources of returns of asset allocation portfolios. Second, many asset classes and sub-asset classes, like a myriad of colors, are mostly combinations of these three primary risk premiums.

On the first point, equity risk premium is the reward to shareholders of company stocks for providing capital to underlying businesses and to the overall economy. Over the years, while there is always debate about its magnitude, the existence of equity risk premium is theoretically sound and empirically proven. On the other hand, the interest rate risk is the reward to bond holders for lending money to governments and corporations over the long term. The upward sloping yield curve, that is, longer-term yields being higher than shorter-term yields, is often a sign of interest rate risk premium. The only risk premium that is somewhat questionable is inflation risk "premium." This is because over the long run, the rate of inflation is roughly in line with risk-free rate, implying little risk premium. However, during the period of high and rising inflation, exposure to inflation risk, such as commodities and inflation-linked bonds, is one of few ways for investor to preserve the real return of their investments. Furthermore, commodities are not a traditional asset class like stocks and bonds. How to think about the role of commodities in a risk parity portfolio is another common question from investors. From a fundamental perspective, one can argue a positive risk premium for investing in commodity futures as a compensation for taking on price risks that commodity producers want to hedge by shorting commodity futures. However, from a technical perspective, commodities as an asset class, offers no dividends or income. Nevertheless, inflation risk "premium" can at least be considered as a hedge for the inflation risk.

If one looks past the names of many asset classes, one would find that they often do not bear any relationship to the risk premiums their names might suggest. Sometime the name, with their usual marketing connotations, could be misleading. The reason is many asset classes are simply hybrid assets. Just like the color purple is a blend of blue and red, hybrid assets represents some blend of equity, interest rate, or inflation risk. The best examples of hybrids are from fixed income. For instance, many asset allocation portfolios would include global government bonds, corporate bonds, HY, emerging market debt, and inflation-linked bonds. But three out of five—corporate bonds, HY, and emerging market debt—have significant exposure to equity risk. In other words, they are not well suited for capturing interest rate premium. On the other hand, inflation-linked bonds' risk exposures are from both interest rate and inflation. The only bond asset class with "pure" exposure to interest rate risk is global government bonds. Even here, some peripheral countries in the Euro-zone have had significant credit risk exposure in recent years prior to the pledge by the European Central Bank (ECB) to "do whatever it takes."

Hence, it is crucial to classify and decompose asset classes to the correct categories of risk premiums. For example, corporate bonds should not be treated as interest rate exposure alone. HY should be classified as equity risk exposure, as we shall demonstrate in an investment insight in this chapter.

Without proper classification of asset classes to risk premiums, an equal risk contribution (ERC) portfolio could be mistaken as risk parity, but it actually has concentrated risk allocation. For example, assume we select four equity asset classes: US equity, non-US equity, small cap equity, and emerging market equity, and the five-bond asset classes above plus commodities. An ERC portfolio with these 10 asset classes would map into a portfolio with a risk allocation of roughly 65% in equity risk, 20% in interest rate risk, and 15% in inflation risk. In addition, when equity and commodity are highly correlated in certain risk on/risk off market environments, the risk allocation could be seen as 80% in risky assets and 20% in safe assets. No wonder this portfolio behaved like a 60/40 portfolio! In summary, a naively constructed ERC portfolio is not necessarily a risk parity portfolio.

We present three investment insights in the remainder of the chapter to address questions related to risk premiums. The first one is on HY, in which we argue HY as an asset class, is "equity in bond's clothing," representing equity risk. The second essay analyzes commodities as an asset class and shows why one should not evaluate it as a traditional asset class like stocks and bonds. In fact, it is rather difficult to evaluate commodities' expected return based on a valuation basis. The last investment insight addresses another common question from investors, do currencies have risk premiums? The answer is no and one should always hedge currency exposure in risk parity or any other asset allocation portfolio.

2.1 HY AS AN ASSET CLASS: EQUITY IN BONDS' CLOTHING*

I never got the midgrade or plus gasoline for my car. I suspect I am not the only driver with this habit—according to the US Energy Information Administration data as of October 2011, out of more than 355 million gallons of gasoline sold and delivered per day, only 13 million was the plus gasoline. The figure for the premium grade was a little higher at 32 million but the majority was in the regular grade, which came in at 309 million or about 87%. What is more interesting and probably unknown is that most gas stations only carry the regular and the premium types but not the plus. When one of the few drivers decides to buy the plus gasoline, the pump simply draws from the regular and the premium at a prescribed proportion such that the mix has the correct octane level of the plus gasoline. You can bet that the mixing is not free.

Here is an analogous investment question: Are HY bonds, as an asset class, the "plus" between investment grade (IG) bonds (the "regular") and equities (the "premium")? Can you mix IG bonds and equities together to produce something like HY? If that is the case, does one need to invest in HY bonds at all?

The answers I arrive at are rather surprising. Yes, at first glance, HY bonds do fall somewhere between IG bonds and equities, in terms of both return volatility and realized Sharpe ratio (excess return over cash divided by excess return volatility). Adding it to a portfolio that holds both IG bond and equities might provide some additional albeit small diversification benefit. However, it is unclear whether the benefit is worthwhile given HY's high correlation to equities, and its illiquidity and tail risk. Upon further study, however, I found that the benefit evaporates completely as we take into account the illiquidity of HY bonds. Measured by the volatility of annual returns instead of the annualized volatility of monthly returns, the risk of HY bonds is about the same as the risk of equities! Over the period from 1984 to 2011, there is very little difference in the risk, return, or Sharpe ratio of HY bonds and equities. In other words, HY bonds are as high octane as equities in terms of return volatility and pro-cyclicality. Furthermore, premium of HY bonds over IG bonds

* Originally written by the author in January 2012.

is highly correlated with the equity risk premium. The evidence strongly suggests that it is more fitting to classify HY bonds as an equity asset class rather than a bond asset class.

2.1.1 Fundamental Linkage between HY Bonds and Equities

HY bonds and equities have a common link to the business cycle. HY bonds are debts issued by companies with a low credit rating, a weak balance sheet, and a high probability of default, compared to IG bonds. In addition, many of these companies are in capital-intensive and cyclical industries, whose prospects depend strongly on the overall economy. When the overall economy is healthy HY issuers can generate sufficient cash flow to cover interest and principal payments and HY bonds generally perform well. However, when the overall economy weakens, many HY issuers encounter difficulty in their underlying business, resulting in a higher probability of default and actual bankruptcy. As a result, HY bonds would perform very poorly. The same dynamics apply to equity shareholders. Equity prices in general depend on the ability of firms to generate profits, which is greatly affected by the business cycle.

Figure 2.1 shows the yield spread between the Barclays US HY Corporate Bond index and the Barclays US Aggregate Bond index. First, notice that the spread spikes up around the recessionary periods in 1990–1991, 2001–2002, and 2007–2008. Second, the spread is generally tight during the expansionary periods from 1994 to 1997, as well as from 2004 to 2007. Third, even during a few brief but dramatic market shocks, such as the 1987 market crash or the LTCM debacle in 1998, the HY spread moved in tandem with the equity market decline.

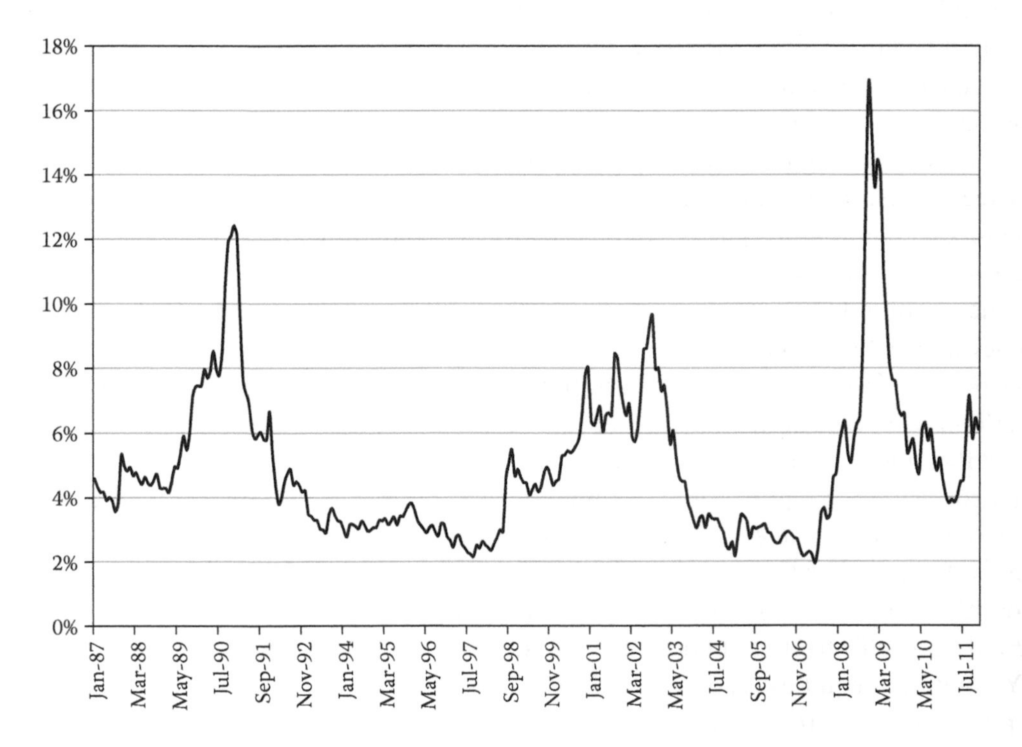

FIGURE 2.1 Yield spread between the Barclays US HY Corporate Bond index and the Barclays US Aggregate Bond index.

Thus, the high correlation between HY bonds and equities is apparent. But are HY bonds different from equities in some way?

2.1.2 A First Glance

A cursory analysis could leave investors with the impression that HY bonds are a hybrid asset class between IG bonds and equities. We present this analysis first. We show later that this analysis is inadequate since it ignores the illiquidity of HY bonds.

We use the Barclays US Aggregate Bond index as a proxy for IG bonds, the Russell 1000 and 2000 indices as proxies for US equities, and the Barclays US Corporate HY index as a proxy for HY bonds. Using monthly returns from January 1984 to December 2011, we calculate excess returns versus the 1-month T-bill, and Table 2.1 shows the annual geometric excess return, annualized volatility of monthly return standard deviation, and the ratio of the two, that is, the Sharpe ratio. Over the period of almost three decades, the return differences are not large, with IG bonds having the smallest return at 3.76% and large cap stocks having the largest return at 5.67%. HY bonds have the second highest return of 4.87%. However, the volatility differences are much larger, ranging from 4.30% for IG bonds to 19.95% for small cap stocks. The volatility of large cap stocks is 15.74% and that of HY bonds is 8.93%, falling nicely in the middle between IG bonds and large cap stocks. The Sharpe ratio is the highest for IG bonds, it is much lower for the equities, and the Sharpe ratio of HY bonds is roughly the average Sharpe ratio of IG bonds and the Russell 1000 index.

The last column shows the statistics of a 50/50 mix in IG bonds and the Russell 1000 index. The numbers are barely distinguishable yet slightly better than those of HY bonds. Indeed, with some hindsight, we have created our own mix for HY bonds using IG bonds and the large cap index. The plus gasoline analogy fits nicely for HY bonds. Or so it seems.

2.1.3 Other Unattractive Features of the HY Bonds

However, the analogy is incomplete in one crucial aspect: our 50/50 mix of IG bonds and the Russell 1000 index (R1) is much superior to HY bonds in terms of liquidity. Fundamentally, individual HY bonds trade very infrequently. There is no public exchange for HY bonds and trading them relies on brokers' ability to source the individual bonds. Their willingness to source them and hold them in inventory varies substantially with market sentiments. When market volatilities spike up, equity markets fall and the liquidity of the HY market tends to dry up, resulting in wide bid-ask spreads and scant trading. In general, when one cannot get in and out of a security efficiently in the short term, the risk of investments increases over the long term.

TABLE 2.1 Annual Excess Return, Annualized Volatility, and the Sharpe ratio of the Four-Asset Classes

	IG Bond	HY Bond	Russell 1000	Russell 2000	50/50 IG/R1
Excess return	3.76%	4.87%	5.67%	4.17%	5.04%
Volatility	4.30%	8.93%	15.74%	19.95%	8.48%
Sharpe ratio	0.88	0.55	0.36	0.21	0.59

Source: The annualized volatility is calculated as square root of 12 times the standard deviation of monthly returns.

TABLE 2.2 Serial Autocorrelation of One Lag, Return Skewness, and Excess Kurtosis, Based on Monthly Returns

	IG Bond	HY Bond	Russell 1000	Russell 2000	50/50 IG/R1
Autocorrelation	0.14	0.34	0.07	0.12	0.06
Skewness	−0.15	−0.81	−0.77	−0.82	−0.55
Excess Kurtosis	0.78	7.67	2.41	3.05	1.50

Source: The returns are from 1984 to 2011.

The consequence of illiquidity is the stickiness of price, which we can measure quantitatively. For fluids like gasoline, one measures the stickiness or resistance to flow by viscosity. In asset prices, one measures illiquidity by the serial autocorrelation of returns. If an asset's liquidity is high, the price adjusts quickly to new information and investors' expectations. As a result, the asset returns tend to be less correlated with prior returns. However, if an asset's liquidity is low, the price only adjusts slowly to incoming information and investors' expectations. This results in a high and positive correlation between consecutive returns.

Table 2.2 displays the autocorrelations with 1-month lag for the four-asset classes, which are all positive. HY bonds have by far the highest autocorrelation at 0.34 and the large cap stocks have the lowest autocorrelation at 0.07 while IG bonds and small cap stocks have a slightly high autocorrelation than large cap stocks. While we cannot attribute everything to illiquidity, the order of autocorrelations agrees with our fundamental intuition regarding the illiquidity of these four-asset classes.

In addition to its exposure to liquidity risk, HY bonds also have excessive tail risk in terms of negative skewness and fat tails. As displayed in Table 2.2, the negative skewness shared by all asset classes and less significant for IG bonds, means the return distribution is skewed to the left or the negative side. The positive excess kurtosis, which is again true for all but less so for IG bonds, indicates the presence of fat tails in the return distribution when compared to a normal distribution. While HY bonds, the Russell 1000 and 2000 indices all share some of these undesirable attributes, it is the combination of negative skewness and high excess kurtosis (7.67) that makes HY bonds most unattractive. In fact, some of the less liquid hedge fund strategies, such as distressed debt and convertible arbitrage exhibit similar return characteristics. In this regard, our 50/50 mix of IG bonds and the Russell 1000 index is much better with a minimal autocorrelation of 0.06 and less tail risk as shown by mild skewness and excess kurtosis.

2.1.4 HY Bonds as an Equity Asset

So far, the data seem to support the notion that HY bonds are a hybrid asset class between IG bonds and equity. Although a naïve 50/50 mix of IG bonds and the Russell 1000 index is superior to HY bonds in terms of risk-adjusted return, better liquidity, and less tail risk, one could still try to make a case that the inclusion of HY bonds may add a potential diversification benefit.

The whole notion hinges on the fact that the volatility of HY bonds is below that of equities. What if that is not true? We now provide two risk measures that show that when the illiquidity of HY bonds is taken into account HY bonds are just as risky as equities.

2.1.5 Annual Return Volatility

First, we use annual returns instead of monthly returns to calculate return volatility. When an asset's liquidity is high, there is little difference in the two approaches. However, when an asset's liquidity is low, one cannot trade or rebalance frequently, the volatility of annual return is a more appropriate indicator of risk since the return variability over the longer period will be much higher than what is suggested by the short-term variability. This alternative method of measuring risk is especially appropriate for long-term strategic asset allocation portfolios.

Table 2.3 shows the same statistics as Table 2.1, with annual return volatility replacing the annualized volatility of monthly returns. First, we note that the annual excess returns are the same as before. But the volatility measures are quite different. While there are some changes for IG bonds and the equity indices, there is a big jump in the volatility of HY bonds, from 8.93% to 16.40% and consequently a big drop in its Sharpe ratio, from 0.55 to 0.30.

This jump in risk, as we move from using annualized monthly standard deviation (times square root of 12) to using standard deviation of annual returns, is true and intuitive. The consequence of high, positive autocorrelation in monthly returns, due to illiquidity, implies that consecutive returns tend to move in the same direction. For example, when return in a month is highly positive, the next month's return tends to be high and positive as well. The opposite is true for negative returns. Hence, a string of monthly returns of the same sign will cause annual returns to have a much wider variation than if the returns are random.*

Now HY bonds appear identical to the equity indices in every statistics: return, risk, and Sharpe ratio. Even though the returns of HY bonds might appear less bumpy than the returns of equity on a short-term basis, they are just as volatile as the equity returns over the long term. In fact, the volatility based on short-term returns could be illusory since one could not trade HY bonds in a sizeable way without having significant price impact. It is as though both are travelers along the same winding path through the mountains of the US Northwest, but one is on a paved new highway while the other is on an alternate old road. They surely experience different ride along the way, but are bounded by the same rise and fall and ultimately reach the same point in the end!

Table 2.3 now clearly shows the 50/50 mix of the IG bonds and the Russell 1000 index is a far better substitute to HY bonds because the annual return volatility of the 50/50 mix is just 9.47%, which is only slightly higher than before.

TABLE 2.3 Annual Excess Return, Annual Return Volatility, and the Sharpe Ratio of the Four-Asset Classes

	IG Bond	HY Bond	Russell 1000	Russell 2000	50/50 IG/R1
Excess return	3.76%	4.87%	5.67%	4.17%	5.04%
Annual volatility	4.72%	16.40%	17.01%	19.14%	9.47%
Sharpe ratio	0.80	0.30	0.33	0.22	0.53

Source: The annual return volatility is calculated as the standard deviation of annual returns from 1984 to 2011.

* By the same token, if the monthly returns have negative autocorrelation or mean-reverting, the annual return volatility tends to be lower than the annualized volatility of month returns.

2.1.6 Liquid HY Bonds

While using annual return volatility corrects the illiquidity bias over a longer horizon, analyzing the index return of liquid HY bonds provides a more direct way to tackle the problem. The Barclays US Corporate Liquid HY index beginning in 1994 includes bonds with better liquidity based on selection criteria in principal outstanding and maturity. We now repeat our analysis above with the liquid HY index and a shorter period from 1994 to 2011.

Table 2.4 shows the annual excess return, annualized volatility, and Sharpe ratio of five assets, including the liquid HY index. It also shows the other three statistics we used to measure illiquidity and tail risks in Table 2.2: autocorrelation with 1-month lag, skewness, and kurtosis. First, we note that the annualized volatility of the liquid HY bonds is 12.07%, which is significantly higher than that of the conventional HY bonds at 9.49%. It is apparent that more liquid HY bonds are more volatile even on the basis of monthly returns than the conventional HY bonds. However, at 12%, the volatility is still lower than that of equities. Second, the Sharpe ratio of the liquid HY bonds is 0.31, which is lower than that of the conventional HY bonds and similar to that of the Russell 1000 index. Both numbers indicate when the illiquidity issue is partly addressed by the liquid index, HY bonds are inching closer to the equity indices and moving away from IG bonds.

However, Table 2.4 also reveals that the liquid HY index has not fully resolved the issue of HY illiquidity. The autocorrelation of the liquid HY index, while declining from 0.31 to 0.2, is still significantly higher than that of the equity indices and IG bonds. The tail risk statistics reflect some improvements over the conventional index but it is still much worse than those of the other assets.

To account for the autocorrelation, we again measure asset risk using standard deviation of annual returns. The resulting results over the period from 1994 to 2011 are presented in

TABLE 2.4 Annual Excess Return, Annualized Volatility, and the Sharpe Ratio

	IG Bond	HY Bond	Liquid HY	Russell 1000	Russell 2000
Excess return	2.98%	4.00%	3.76%	4.48%	4.11%
Annualized volatility	3.75%	9.49%	12.07%	15.86%	20.34%
Sharpe ratio	0.79	0.42	0.31	0.28	0.20
Autocorrelation	0.14	0.31	0.20	0.11	0.10
Skewness	−0.22	−0.94	−0.64	−0.66	−0.47
Excess kurtosis	0.83	8.11	5.33	1.09	0.95

Source: The annualized volatility is calculated as square root of 12 times the standard deviation of monthly returns.

TABLE 2.5 Annual Excess Return, Annual Return Volatility, and the Sharpe Ratio of the Four-Asset Classes

	IG Bond	HY Bond	Liquid HY	Russell 1000	Russell 2000
Excess return	2.98%	4.00%	3.76%	4.48%	4.11%
Annual volatility	4.68%	17.50%	19.48%	19.42%	19.39%
Sharpe ratio	0.64	0.23	0.19	0.23	0.21

Source: The annual return volatility is calculated as the standard deviation of annual returns from 1994 to 2011.

TABLE 2.6 Correlations of Annual Returns from 1984 to 2011

	IG Bond	HY Bond	Russell 1000	Russell 2000	HY less IG
IG Bond	1.00	0.36	0.22	0.22	0.07
HY Bond	0.36	1.00	0.70	0.80	0.96
Russell 1000	0.22	0.70	1.00	0.84	0.68
Russell 2000	0.22	0.80	0.84	1.00	0.79
HY less IG	0.07	0.96	0.68	0.79	1.00

Table 2.5 for the fives assets. The same pattern emerges: HY bonds (both the liquid index and the conventional index) and equities are barely indistinguishable from one another with similar return, risk, and Sharpe ratio. Note the annual return volatility of the liquid HY index is essentially of the same level of the two equity indices!

2.1.7 Correlation Analysis

It is apparent that the return and risk of the HY bonds for the last three decades are no different from those of equities. What about correlation?

Table 2.6 displays the correlation matrix of annual returns from 1984 to 2011. Also included is the correlation of the HY premium over the IG bonds with the four-asset classes. Among the four-asset classes, HY bonds do have a slightly higher correlation with IG bonds (at 0.36) than the two equity asset classes (both at 0.22). But HY bonds have much higher correlations with the equity indices. The correlation with the Russell 1000 index is 0.70 and the correlation with the Russell 2000 index is 0.80, both of which is only slightly lower than the correlation between the Russell 1000 index and 2000 index (at 0.84). It is clear that from the correlation perspective, the HY bonds fit well with equities as well.

The correlations of the HY premium over IG bonds tell the same story. The premium correlates highly with HY bonds and the equity indices and is uncorrelated with the returns of IG bond. It is evident that the HY premium and the equity risk premium are two of the same kind of growth risk—they rise and fall together. It would be mistaken to think of them as two independent sources of risks.

2.1.8 Conclusion

HY bonds and equities have a lot in common. They are both greatly influenced by the business cycle and they are both lowly ranked in the pecking order of corporate stakeholders. Despite the strong commonality, the conventional wisdom is that HY bonds are a hybrid asset class between IG bonds and equities. This is partly because the naming of HY bonds creates a simple yet false classification with respect to the asset class to which it belongs. In addition, it is based on the notion that its risk level is lower than that of equities. We find this to be false too. After accounting for the illiquidity of HY bonds, either by using volatility of annual returns or the liquid HY index, the volatility of HY bonds is as high as the volatility of equities. Over almost three decades from 1984 to 2011, there is virtually no difference between HY bonds and large cap and small cap equities, in terms of excess return, risk, and Sharpe ratio. Indeed, my initial analogy of the plus gasoline is not appropriate for HY bonds. A more appropriate assessment is that HY bonds are equity in bond's clothing.

This realization has several clear implications to asset allocation decisions. First, one should properly calibrate the risk inputs regarding HY bonds by using volatility of annual returns. Second, in deciding the allocation between high-risk growth assets and low-risk defensive assets, put HY bonds in the high-risk growth category where they rightly belong. Third, equity portfolios should have the freedom to invest in HY bonds and likewise bond portfolio should have the freedom to invest in equities in lieu of HY bonds. Fourth, one must realize that equity risk premium and the credit premium of HY bonds are one and the same. In the context of risk parity portfolios, allocating equal risk budget to both equity and credit amounts to potentially doubling down on equity risk. By all accounts, due to its equity exposure and illiquidity, the case can be made forcefully that HY bonds should not be included in risk parity portfolios.

2.2 ROLL YIELDS, PRICES, AND COMMODITY RETURNS*

One recent evening my third-grade daughter wanted to do some math. Seizing the opportunity, I opened a math problem book for elementary and middle-school students. After flipping through the pages, I found one: "A girl bought a dog for $10, sold it for $15, bought it back for $20, and finally sold it for $25. Did she make or lose money and how much did she make or lose?"

After some initial disappointment about not getting a dog, she started thinking. "She made $10." "How did you figure that out?" "She made $5 first and then she made another $5. So that's $10." I praised her for getting the right answer and she was happy.

Strangely, the problem and my daughter's solution left me unsatisfied and my thoughts went back to something that had been on my mind for quite some time. I could not help but noticing there is a similarity between this problem and commodity investing. In commodity terminology, the girl had a negative roll of $5 when she sold it for $15 and bought it back for $20. Yet my daughter did not have such a notion to get the right answer. In third grade math, profit is generated from buying low and selling high regardless of roll yield. So why in the investment management world would investors care so much about roll yield when investing in commodities? Is roll yield actually a yield in the traditional sense? Is roll yield important in determining long-term returns in commodities? What are the risk premiums of commodities? These are very important questions as global investors diversify their portfolios to include alternative investments.

In fact, there is a lot of confusion regarding these questions. One point of confusion is equating roll yield as a valuation measure. Another related mistake is to regard roll yield as a determining factor in long-term commodity returns (Inker, 2011). Both claims are dubious. I shall try to clarify both points in this research note.

2.2.1 Roll Yield and Third-Grade Math

Commodity investing often entails investing in commodity futures. Since futures expire, investors have to roll the contracts forward. For example, an investor with a long position has to sell the maturing contract (the front contract) and buy futures with later maturing

* Originally written by the author in November 2011.

dates (the back contract). The roll return or roll yield concerns the price difference between the front and back contracts. If the price of the back contract is higher than the price of the front contract, there is a negative roll yield and the market is in contango, like the price of the dog in the third-grade math problem. On the other hand, if the back contract price is lower than the price of the front contract, the roll yield is positive and the market is considered to be in backwardation.

The first question to address is why the roll yield is relevant at all. In the dog problem, according to my third grader, it is not. Here is how she (probably everyone including commodity traders) came up with the answer:

$$(\$15 - \$10) + (\$25 - \$20) = \$5 + \$5 = \$10.$$

This makes perfect sense: you calculate profit and loss on the two pairs of buy and sell transactions. The same procedure could apply to commodity futures. For instance, if one bought and sold two commodity contracts sequentially, one first calculates the profit and loss on the two contracts, respectively, and then sums them up.

Apparently, this solution is too simple to satisfy commodity traders. They solve the problem another way by including the impact of the negative roll of \$5:

$$-\$10 + (\$15 - \$20) + \$25 = -\$10 + (-\$5) + \$25 = \$10.$$

Obviously, the two sets of calculations lead to the same answer. The only difference is the sequence of operations. However, comparing two equations does reveal a very important point: if one includes roll returns in calculating profit and loss, one must also consider *price changes*!

In the dog problem, the ending price of \$25 is higher than the initial price of \$10 by \$15, which more than offsets the negative roll of \$5, leaving the girl with a \$10 profit. Suppose the ending price had stayed at \$10, the loss would be \$5, equal to the negative roll. The point is that "the roll yield is important only if the price change is small compared to the roll yield." Extending the argument to cases where there are multiple rolls over multiple periods, the profit and loss will consist of two parts: one is the cumulative return of the roll yield and the other is the price difference of the initial contract and the ending contract. The relative magnitudes of the two determine their relative impact on the overall returns.

2.2.2 Roll Yield is Not Value

It is a tautology that in any investment both yield and price return matter. The name "roll yield" certainly evokes the idea of valuation. Is roll yield similar to traditional valuation measures such as bond yield and dividend yield? The answer is no.

First, we note that unlike coupon interest or dividends, roll yield is not actual income to investors because commodities pay neither a coupon nor a dividend. If one invests in physical commodities rather than futures, the only source of return is price appreciation, after accounting for storage costs. So, if there is no income yield in physical commodities, why would it exist in the futures market? This has to do with how futures are priced and modeled.

For financial futures, such as equity index or treasury bond futures, a futures price F is determined by a spot price S, the short-term interest rate r, the dividend yield c, and the time to maturity $(T-t)$ through the exponential function:

$$F = Se^{(r-c)(T-t)}.$$

If the short-term interest rate is higher than the dividend yield, that is, $r-c>0$, the futures price is higher than the spot price, or in contango. On the other hand, if the dividend yield is higher than the short-term interest rate, that is, $r-c<0$, the futures price will be lower than the spot price, or in backwardation. This futures-spot parity relationship works well for financial futures, because all the inputs on the right-hand side of the equation are known and arbitrage strategies can be deployed to reduce meaningful deviations from equilibrium.

Unfortunately, this framework does not fully apply to commodity futures because it could never justify a backwardated commodity futures curve because $r-c$ is always greater than zero when the yield is zero. The fix to the framework is to replace dividend by the so-called convenience yield in the formula (this is why I used c instead of d) so that $r-c$ could be less than zero. The economic intuition of convenience yield is that when commodity supplies are in shortage, it is more advantageous or convenient to hold physical commodities. As a result, a high convenience yield would lead to a backwardated futures curve. On the other hand, when supplies are abundant, it is more advantageous to hold futures instead of physicals. The convenience yield would be negative (or an inconvenience yield) justifying a futures curve in contango.

The concept of the convenience yield appears to have been constructed in order to fit theoretical pricing models to explain the demand and supply imbalances of commodity markets. Another theory involves the demand and supply of hedgers (producers and users) and speculators. Regardless of the theory, they are not the same as the traditional valuation measures as bond yield and dividend yield.

The second difference between the roll yield and traditional valuation metrics is that the roll yield can and often does change dramatically over a short time period. Seasonal patterns related to weather and harvest, commonly impacts a commodity's roll yield. In contrast, bond yields or dividend yields are almost always positive and tend to be very persistent over time. The fact that the roll yield can swing widely and change signs disqualifies it as a reliable measure of long-term valuation.

A third difference is more subtle and perhaps even counterintuitive. Traditional valuation measures typically rise with falling prices and vice versa because they are expressed as ratios of income to asset prices. However, because the roll yield is the ratio of two prices, it could actually rise when commodity prices rise. For instance, when there is a negative supply shock, commodity prices rise and the futures curves typically flatten, implying higher roll yields. From this perspective, as prices get expensive, "values," as measured by roll yield, paradoxically get cheaper! Active managers often use value and momentum factors in their strategies. In a sense, "value" and momentum become the same in commodity markets. Based on traditional valuation metrics in other asset markets, this phenomenon is rare since value and momentum are in general negatively correlated.

In some sense, as demonstrated in the dog math problem, the roll yield is just an accounting plug with respect to futures returns. It is not a traditional value metrics. Rather, it is a short-term signal to determine supply and demand imbalances and consequently is better suited as a signal for short-term tactical trading strategies. For example, it is well documented that relative roll yield (or basis) is a predicting factor for relative future returns over a time horizon of a few months (Gorton and Rouwenhorst, 2006). As previously shared, roll yield is a useful predictor of futures returns when it is stable and it is large relative to price changes. This is certainly true in general when both roll yield and price changes are measured over a short-term horizon and on a relative basis.

2.2.3 Return Attribution

Nevertheless, as an attribution exercise, one can still attribute investment returns in commodities to roll yield and price return, and analyze their respective contributions. In the rest of the note, we shall carry out attribution analysis for select commodities, and for a cross-sectional selection of commodities over multiple periods, covering 1970 to 2010. Our analysis confirms what we have learned from our simple example: when the commodity prices are stable over time, the roll yield is an important factor in determining commodity returns. Conversely, when commodity prices change dramatically, either due to changes in inflationary pressure or economic demand, the roll yield is rendered irrelevant and price returns dominate.

2.2.4 Gold and Copper

First, we note that negative roll yields are not necessarily detrimental to futures returns. A case in point is precious metals such as gold. Because gold, in addition to having no income, incurs little storage cost compared to other commodities, its futures price curve is almost in permanent contango—the price of the front contract is lower than the price of the back contract by an amount related to the short-term interest rate. As a result, every time investors roll a gold futures contract, there is a negative roll yield. Another example is equity index futures, whose prices are affected by short-term interest rates as well as dividend yields. When the former is higher than the latter (not true today since short-term rates for many countries are near zero, lower than dividend yields), the index futures will be in contango. Obviously, the fact that equity index futures or gold futures might have a negative roll yield has not deterred many investors from buying them to gain long exposures to these investments. In both cases, changes in prices, which is partly driven by profit growth in the case of equities and inflationary or currency debasement concerns in the case of gold, determine the investment outcomes.

Figure 2.2 plots the cumulative return of the front gold futures contract and the cumulative return of its roll yield, from 1975 to September 2011. Also plotted is the price of the front futures contract (scale on the right). It is apparent that the two cumulative return lines bear little relationship to each other. Since the roll yield is always negative, tied to the US short-term interest rate, the cumulative line for the roll yield is a smooth and declining curve. On the other hand, the actual futures return is much more volatile. We can roughly divide the sample into three periods. The return was strong in the late 1970s driven by

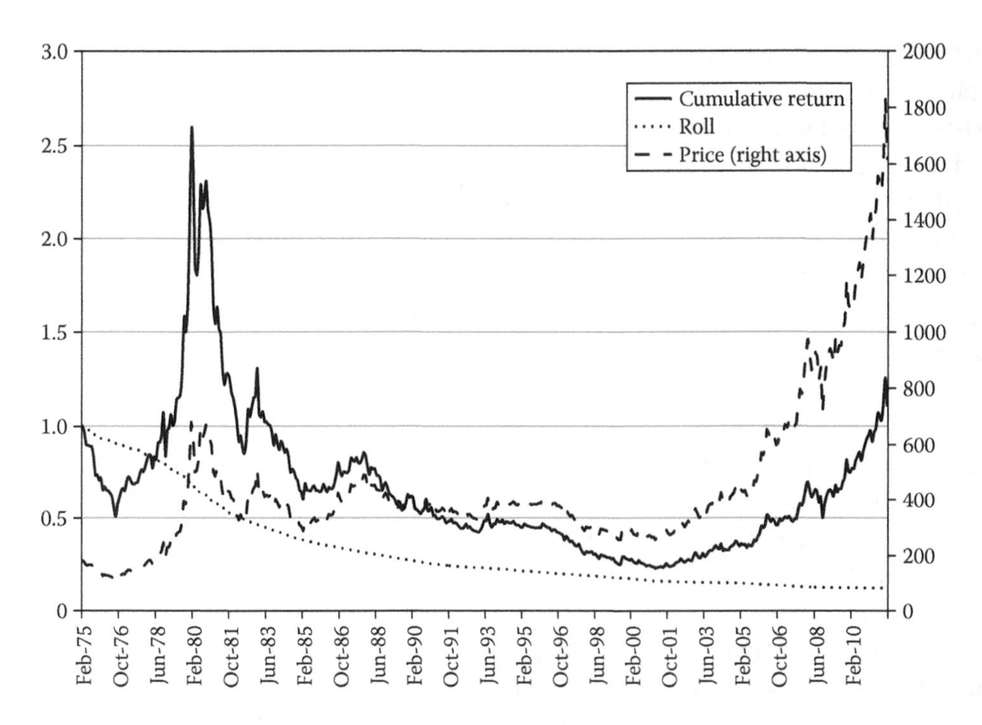

FIGURE 2.2 Cumulative return of gold futures and roll and futures price.

rising prices. After the price of gold collapsed, the price became much more stable from 1990 to 2001. During this period, the negative roll yield contributed more meaningfully to the return of gold. Starting in 2001, the price of gold rose significantly, bringing the return with it. Therefore, the only period during which the roll yield is important in terms of return contribution is from 1990 to 2001 when the price of gold exhibited low volatility.

Table 2.7 shows the annual averages and standard deviations of the futures return as well as the roll yield, correlation, and R-squared of the linear regression of the futures return versus the roll yield. The volatility of actual returns is much higher than the volatility of roll yield. The correlation is very low. Over time, the roll yield explains little in the variation of the actual return of gold. Statistically, it is evident that the roll yield is irrelevant for gold.

How about other commodities, whose roll yields are greatly influenced by supply and demand? To test this we use copper because of its broad use in construction, electronics, infrastructure, and transportation. Is the roll yield important for copper? The answer is ambiguous at best.

TABLE 2.7 Annual Return Statistics for Gold Futures and Roll Yield

	Return	Roll Yield
Average	0.9%	–5.6%
Standard deviation	25.0%	3.2%
Correlation		0.1
R^2		0.0

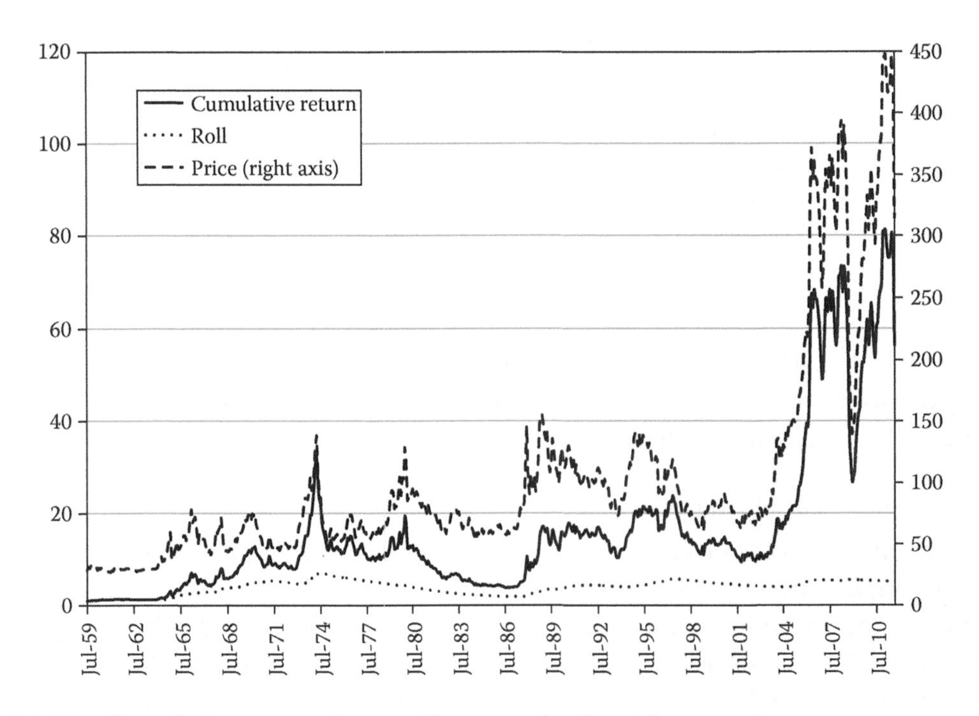

FIGURE 2.3 Cumulative return of copper futures and roll and futures price.

Figure 2.3 plots the cumulative return from investing in the front copper futures contract as well as the cumulative return of copper's roll yield from 1959 to September 2011. Also plotted is the price of the front copper futures contract (scale on the right). We again notice that the return is much more volatile than the roll yield. The most noticeable feature of the graph is the rapid rise in the price and significant positive return since 2001 while the curve for the roll yield stayed flat. During the previous years, the price was relatively stable and the roll yield was more relevant.

Table 2.8 presents the annual average and standard deviation for the return and the roll yield, as well as the correlation and R-squared of the regression. The roll yield for copper has a much higher volatility at 15.4%. And its correlation with the futures return is 0.5 and over time it explains 25% of the variance of futures returns.

A close look at subperiods shows that the importance of copper's roll yield has changed over time. Table 2.9 shows the annual average of the futures return, the roll yield, and the price return for the past five decades. Prior to the last decade, the magnitudes of the futures return and the roll yield were comparable. However, from 2001 to 2010, the roll

TABLE 2.8 Annual Return Statistics for Copper Futures and Roll Yield

	Return	Roll Yield
Average	8.6%	3.2%
Standard deviation	48.2%	15.4%
Correlation		0.5
R^2		0.25

TABLE 2.9 Average Returns for Futures and Roll

	Futures Return (%)	Roll Yield (%)	Price Return (%)
1960–1970	19.92	15.78	3.58
1971–1980	2.23	−3.12	5.53
1981–1990	4.76	1.20	3.51
1991–2000	−2.44	0.88	−3.29
2001–2010	19.49	1.24	18.02

yield averaged 1.24% per year while the futures return averaged close to 20%, which is mostly due to the price return. We also note that only during the period from 1960 to 1970 was the roll yield meaningfully positive.

2.2.5 Cross-Sectional Analysis

After studying the role of roll yield for gold and copper, we present cross-sectional results across multiple commodities and across multiple periods. The results cover commodities in precious metals, industrial metals, energy, grains, live stocks, and softs. The period is from 1971 to 2010, which is divided into four decades. The results show that the roll yield was a minor factor in determining futures returns. It is the price return that was the most dominant contributing factor.

Let us start with the most recent decade. Table 2.10 displays the annualized return of futures, roll yield, and price for 13 commodities during the 10-year period from 2001 to 2010. As seen from Table 2.10, the price appreciation experienced by gold and copper was not unique for this period. All commodities had positive price returns, many of them in double digits, with an annualized average of 11.8%. However, the roll yields averaged −4.4% during the period. The futures returns were significantly higher than the roll yield with the

TABLE 2.10 Return Attribution of Select Commodity Futures Returns from 2001 to 2010

2001–2010	Futures Return (%)	Roll Yield (%)	Price Return (%)
Wheat	−5.0	−14.4	11.0
Soybean	13.1	2.0	10.8
Corn	−6.0	−14.9	10.5
Live cattle	6.6	3.5	3.0
Cotton	−5.0	−12.7	8.8
Coffee	−2.4	−4.1	1.7
Sugar	10.3	−1.7	12.2
Cocoa	15.3	0.4	14.9
Gold	14.6	−2.8	17.9
Silver	17.4	−3.0	21.0
Crude oil	6.9	−5.5	13.1
Heating oil	4.3	−5.9	10.9
Copper	19.5	1.2	18.0
Average	6.9	−4.4	11.8
Standard deviation	9.1	6.1	5.5

average being 6.9%. The commodities with the highest return were copper, silver, gold, cocoa, and soybean, while wheat, corn, cotton, and coffee had negative returns. In summary, during this 10-year period, the actual futures returns were significantly higher than the roll yields. The plausible reasons for the price increase include a weakening US dollar (USD), demand from emerging markets, and investor demand for alternative investments.

Another decade that most resembles the recent one is the 1970s. Table 2.11 shows the return attribution for a smaller set of commodities for which we have futures returns. The price returns were high during this period, while the roll yields were again generally negative with the exception of coffee and cocoa. The futures returns were all positive with an annualized average of 9.9%.

In summary, for these two decades, roll yields had negative contributions to futures returns while price had positive contribution. This divergence could never happen for traditional assets like stocks and bonds, to which the contributions of dividend and coupon interest are always positive and often significant. For example, based on our research, coupon interest income usually accounts for 75% to 100% of fixed-income total returns.

The two decades from 1981 to 2000 saw a role reversal, where the roll yields were just as important as the price return in determining the futures returns, since the commodity prices were more stationary during those two decades. As seen from Table 2.12, from 1981 to 1990, the average roll yield was −2.6% and the average price return was −3.6%. Thus, in aggregate, both contributed to the negative futures returns. There were some large variations across different commodities, however. For example, the roll yield for heating oil was very high at 21%, making up the majority of the futures return. On the other hand, the price return of coffee was positive 3.8% while the roll yield was negative. For many other commodities, both roll yields and price returns were negative and of comparable magnitude. Also note that the standard deviation of price returns was rather small at 5.1% while that of the roll yields was much higher.

TABLE 2.11 Return Attribution of Select Commodity Futures Returns from 1971 to 1980

1971–1980	Futures Return (%)	Roll Yield (%)	Price Return (%)
Wheat	7.7	−3.4	11.5
Soybean	8.4	−1.8	10.4
Corn	2.2	−6.5	9.3
Live cattle	4.2	−4.4	9.0
Cotton	12.6	−0.9	13.6
Coffee (1973)	16.3	6.2	9.6
Sugar	4.9	−13.7	21.5
Cocoa	30.3	15.6	12.7
Gold			
Silver			
Crude oil			
Heating oil			
Copper	2.2	−3.1	5.5
Average	9.9	−1.4	11.5
Standard deviation	9.0	8.2	4.4

TABLE 2.12 Return Attribution of Select Commodity Futures Returns from 1981 to 1990

1981–1990	Futures Return (%)	Roll Yield (%)	Price Return (%)
Wheat	−11.2	−5.2	−6.3
Soybean	−10.4	−7.2	−3.4
Corn	−9.1	−4.5	−4.8
Live cattle	8.3	6.9	1.3
Cotton	6.4	8.6	−2.0
Coffee	−1.7	−5.3	3.8
Sugar	−31.2	−22.5	−11.2
Cocoa	−10.3	−5.0	−5.6
Gold	−11.8	−8.0	−4.1
Silver	−21.7	−10.7	−12.4
Crude oil			
Heating oil	18.9	21.0	−1.7
Copper	4.8	1.2	3.5
Average	−5.8	−2.6	−3.6
Standard deviation	13.8	11.0	5.1

Finally, the decade from 1991 to 2000 shows the same pattern with an even stronger influence of roll yields on futures returns. As seen from Table 2.13, the price returns were rather small for all commodities, with a minimum of −4.1% for cocoa and maximum of 2.3% for copper. The cross-sectional average and standard deviation were merely −0.4% and 1.9%, respectively. As a result, the roll yields accounted for the majority of futures returns for most commodities. The roll yield was positive for crude oil, heating oil, copper, sugar, and live cattle while it was negative for the other, mostly agriculture, commodities. Overall, the futures returns follow the roll yield rather closely. Among the four decades

TABLE 2.13 Return Attribution of Select Commodity Futures Returns from 1991 to 2000

1991–2000	Futures Return (%)	Roll Yield (%)	Price Return (%)
Wheat	−6.9	−7.5	0.7
Soybean	−3.1	−2.0	−1.1
Corn	−7.9	−7.9	0
Live cattle	2.6	2.2	0.4
Cotton	−4.8	−2.6	−2.2
Coffee	−7.8	−8.5	0.7
Sugar	6.6	5.7	0.9
Cocoa	−15.8	−12.2	−4.1
Gold	−7.7	−4.2	−3.6
Silver	−4.1	−4.9	0.9
Crude oil	6.4	7	−0.6
Heating oil	10.9	9.7	1.1
Copper	6.1	3.7	2.3
Average	−1.9	−1.7	−0.4
Standard deviation	7.8	6.8	1.9

we studied, roll yields offered a significant contribution to futures returns only during the period from 1991 to 2000.

2.2.6 Conclusion

This research note dispels some of the misconceptions regarding roll yields and commodity returns. First, we argue that roll yield is not a traditional valuation metric. As a result, it should not be used to forecast long-term returns of commodity futures. Empirically, our results show that during the last 40 years the return of most commodities is dominated by changes in price. While it is very challenging to estimate long-term returns of stocks and bonds, it is much harder to estimate long-term return premium of commodities due to the lack of a positive valuation anchor. Therefore, one should use extra caution in estimating long-term returns of commodities.

Aside from tactical trading, investors should not base investment decisions regarding commodities on merely capturing the roll yields, which is not a good predictor of return premium. Rather, in our view, one should base strategic asset allocation decision on two criteria: one is diversification and the other is return premium stemmed from strong demand. First, while the current macroeconomic environment with weak growth in the developed world is not conducive to higher inflation, maintaining an allocation to real assets such as commodities can provide a hedge against future inflation shocks, which would have a negative impact on traditional assets such as nominal bonds and stocks. Second, the demand for commodities from emerging market economies could continue to grow. That could lead to a further price increase in some commodities such as oil and industrial metals. Based on these considerations, we believe investments in commodities provide investors both diversification benefits as well as the potential for additional returns.

2.3 DO CURRENCIES HAVE RISK PREMIUMS?*

One of the questions from investors regarding risk parity portfolios is whether one should have an allocation to currency risk. The implication is that, like equity risk or interest rate risk, currency risk might also provide a risk premium to investors and therefore deserves a place in a risk parity portfolio.

I have been somewhat surprised by this question since it seems apparent to me that, currencies do not have inherent risk premiums like equities or interest rates. So was my belief wrong or have some investors confused "currency carry factor premium" with currency risk premium?

In this research note, I provide both analytical and empirical evidence that show currency risk has no return premium. While exposure to currency carry (long higher yielding currencies and shorting lower yielding currencies) might offer a return premium, it is not the same as equity or interest rate risk premium. However, there may be a reason for some investors in certain base currencies to consider including currency risk in their portfolios. This motivation originates from the perspective of risk management rather than return capturing, as foreign currency exposure can at times serves as a useful hedge to

* Originally written by the author in January 2014.

their portfolio. This hedging is most useful during a time of financial stress but it seems to come at a cost to the portfolio's return over the long run.

2.3.1 Risk Premium 101

Risk premium, or the expected return of an investment in excess of the return on the risk-free asset or cash, is a required return to compensate investors for taking investment risk. Not all risks have rewards or risk premiums.

First, it is useful to consider why some risks have risk premiums from a fundamental perspective. Take equity risk premium, for example. One can think of it as compensation to investors for bearing the business risk of companies. Shareholders provide capital to the company to engage business activities, which could be quite risky. If the business is successful, the shareholders would be rewarded with a higher share price and larger dividends in the future. But if the business is not successful or fails, the shareholder could suffer losses. For the economy as a whole, equity risk premium is the aggregated return to investors for bearing economic growth risk.

By the same token, interest rate premium is compensation to investors for bearing inflation risk that could erode their future purchasing power. If future inflation is lower than expected, bond yields might fall or rise less than what has been priced in. But if future inflation is higher than expected, bond yields might increase more than expected. Interest rate premium depends on the bond's maturity—the longer the maturity, the higher the premium to compensate for the higher uncertainty of inflation risks.

Second, from an empirical perspective, evidence for equity risk premium and interest rate premium has been very strong. In Table 2.14, we list the excess returns and annualized standard deviations of three-asset classes: 5-year US government bond, 10-year US government bond, and the S&P 500 index, from 1919 to 2013. The excess returns are 1.31%, 1.50%, and 6.31% for the 5-year bond, 10-year bond, and the S&P 500 index, respectively. The risk, or return volatility, is low for US bonds and much higher for equities—it is 4.56%, 6.53%, and 18.77% for the 5-year bond, 10-year bond, and the S&P 500 index, respectively. As a result, the Sharpe ratios are comparable, at 0.29, 0.23, and 0.34. With such a long data history, the t-stats for the excess returns of all assets are very significant.

2.3.2 No Reason for Currency Risk Premium

Now let us similarly examine currency risk from both a fundamental and an empirical perspective. First, it is hard to see any fundamental reason why foreign currency exposure would bring excess return relative to the domestic risk-free rate. Suppose, instead of depositing currency in a domestic bank, one converts the currency into foreign currency and

TABLE 2.14 Statistics of Risk Premiums of Interest Rates and Equity

	5-year	10-year	S&P 500
Excess return	1.31%	1.50%	6.31%
Standard deviation	4.56%	6.53%	18.77%
Sharpe ratio	0.29	0.23	0.34
t-stat	9.69	7.75	11.34

deposits the proceeds in a foreign bank. The deposit is now earning the foreign risk-free rate. However, it is not put into any productive use. The bank might lend it out to its customers but any profit from that transaction is for the bank to keep. Yes, the investor's capital is now subject to exchange rate fluctuation or currency risk; the return is equal to the sum of the foreign risk-free rate and the foreign currency return. But it is hard to fathom any fundamental reason why this return should be better than the local risk-free rate.

Even on a more basic level, currency risk premium could not possibly exist for all investors with different base currencies, because a gain for foreign currency exposures to some investors must be a loss to other investors with the opposite foreign currency exposures. It is indeed a zero-sum game—currency exposures cannot bring all positive (or all negative) premiums to all investors.

This is fundamentally different from equity risk premium and interest rate premium, which are available to all investors. Investors can gain these risk premiums from both local and foreign markets, *if foreign investments are currency-hedged*. For example, US investors get the excess return of US equities and they also get the excess return of UK equities (vs. UK risk-free asset) by investing in UK equities while hedging the British Pound back to the USD. The latter is the same return UK investors would earn from investing in UK equities. In other words, equity risk premiums from different countries are available to all investors with different base currencies as long as the investments are currency-hedged.

2.3.3 Where Is Currency Risk Premium?

There are many ways to calculate currency returns, in terms of currency pairs, or indices with a basket of currency exposures. In this note, we use differences between hedged and unhedged equity and government bond return indices to deduce any potential premium from currency risk that are in the unhedged indices but are absent in the hedged indices. The benefit of this approach is it allows us to clearly distinguish equity and interest rate premiums from currency premium.

We first consider the Citigroup World Government Bond Index (WGBI) from both a hedged and unhedged basis, with several different base currencies. The monthly return data cover the period from February 1985 to December 2013.

Table 2.15 shows the returns to USD-based investors. On a hedged basis, the annualized return is 7.01% and the risk is 3.43%, with a return t-stat of 3.1. When the index is unhedged, which carries exposure to a basket of foreign currency risk, the annualized return is 7.90% and the risk increases dramatically to 7.06%. The last column denotes the difference between the unhedged and hedged indices, or currency surprise; it has an annualized return of 0.86% and a risk of 5.61%. In this case, the unhedged index introduces

TABLE 2.15 WGBI Return Statistics for USD Investors

	USD Hedged	USD Unhedged	Currency
Return	7.01%	7.90%	0.86%
Volatility	3.43%	7.06%	5.61%
t-stat	3.10	1.73	0.28

substantial currency risk, which is much higher than interest rate risk in the hedged index, but with minimal return. The t-stat of the average return is a meager 0.28.

The results from the perspective of other currencies are similar. Tables 2.16 through 2.19 display return statistics for four additional currency perspectives: British Pound (GBP), Japanese Yen (JPY), Euro (EUR), and Australian dollar (AUD), respectively. In all cases, the hedged returns are significantly positive, while the unhedged returns are not very different but have significantly higher risks. The currency risks in each case have no meaningful premium. As we argued previously, the returns to currency exposures show mixed signs, positive for some and negative for others.

A few specific features are worth noting. First, the hedged return is relatively low for the JPY-based investors while it is relatively high for Australian investors. This is due to the low risk-free rate in Japan and the high risk-free rate in Australia. However, the difference between excess returns, relative to risk-free rates of respective base currency should be smaller. Second, currency return volatility varies across different base currencies. This is due to both pairwise currency volatilities and the weights of foreign markets in the index. For example, the WGBI index has significant weight in the United States and JGBs but little weight in Australian government bonds. Australian investors investing in an unhedged

TABLE 2.16 WGBI Return Statistics for GBP Investors

	GBP Hedged	GBP Unhedged	Currency
Return	9.14%	6.48%	−2.46%
Volatility	3.55%	8.77%	7.94%
t-stat	3.87	1.18	(0.42)

TABLE 2.17 WGBI Return Statistics for JPY Investors

	JPY Hedged	JPY Unhedged	Currency
Return	4.25%	4.64%	0.33%
Volatility	3.42%	8.18%	8.06%
t-stat	1.92	0.93	0.13

TABLE 2.18 WGBI Return Statistics for EUR Investors

	EUR Hedged	EUR Unhedged	Currency
Return	7.51%	5.41%	−1.98%
Volatility	3.52%	7.17%	6.54%
t-stat	3.23	1.20	(0.42)

TABLE 2.19 WGBI Return Statistics for AUD Investors

	AUD Hedged	AUD Unhedged	Currency
Return	10.53%	7.54%	−2.69%
Volatility	3.58%	12.32%	11.43%
t-stat	4.39	1.01	0.28

WGBI would be exposed to a greater level of currency risk than both US and Japanese investors would. Third, Australian investors had the most negative return from currency exposure during this period. We shall return to this fact later in the note.

2.3.4 Keep Searching for Currency Risk Premiums

Similar results are expected when we compare hedged and unhedged equity return indices, since currency risk exposure is based on the same underlying currencies as in the bond case. The only difference is the weights of the currency exposures are different.

It is nevertheless informative to show these results because they reveal the relative magnitudes of equity risk and currency risk and, more importantly, the correlation between equity risk and currency risk for different base currencies.

Table 2.20 shows the return statistics of the Morgan Stanley Capital International (MSCI) World Index for USD investors. The data goes from January 1988 to December 2013. It is noted the hedged and unhedged indices have virtually the same return and comparable risk. As a result, the currency risk has no return. The correlation between the hedged return series and currency return is slightly positive. This is why the volatility of the unhedged return is higher than the volatility of the hedged returns. We also note that for this period, the equity return is quite low, both absolutely and relatively versus the WGBI indices. This results in a low t-stat for the equity returns.

The results from the other four currency perspectives are largely similar, as shown in Tables 2.21 through 2.24. The currency returns have mixed signs. Comparing Tables 2.20 through 2.24 to Tables 2.15 through 2.19, we note the signs of currency returns are the same for the equity indices and bond indices. None of them is statistically significant, however.

The correlation between currency returns and hedged equity returns are quite different for different currencies. For Japanese investors, it is positive 0.15. This indicates that when equity returns are positive, currency returns tend to be positive as well, that is, JPY tends to depreciate against other currencies in strong equity markets. On the other hand, when equity returns are negative, currency returns tend to be negative too, as the JPY broadly appreciates against other currencies. This positive correlation results in a higher volatility

TABLE 2.20 MSCI Return Statistics for USD Investors

	USD Hedged	USD Unhedged	Currency
Return	5.56%	5.55%	0.04%
Volatility	14.14%	15.18%	4.50%
t-stat	0.67	0.64	0.05
Corr (H,C)			0.08

TABLE 2.21 MSCI Return Statistics for GBP Investors

	GBP Hedged	GBP Unhedged	Currency
Return	7.29%	6.08%	−1.22%
Volatility	14.16%	15.28%	7.08%
t-stat	0.84	0.68	0.20
Corr (H,C)			−0.09

TABLE 2.22 MSCI Return Statistics for JPY Investors

	JPY Hedged	JPY Unhedged	Currency
Return	3.00%	4.98%	2.13%
Volatility	14.02%	17.53%	8.63%
t-stat	0.41	0.54	0.42
Corr (H,C)			0.15

TABLE 2.23 MSCI Return Statistics for EUR Investors

	EUR Hedged	EUR Unhedged	Currency
Return	1.35%	1.04%	−0.59%
Volatility	14.81%	14.81%	7.36%
t-stat	0.18	0.16	0.05
Corr (H,C)			−0.25

TABLE 2.24 MSCI Return Statistics for AUD Investors

	AUD Hedged	AUD Unhedged	Currency
Return	7.59%	4.70%	−3.38%
Volatility	14.32%	13.55%	10.72%
t-stat	0.86	0.60	0.39
Corr (H,C)			−0.44

of unhedged MSCI returns from a JPY perspective. However, in this case the currency exposure also brings a return of about 2%.

This correlation dynamic is completely different for Australian investors. The correlation between the hedged equity return and the currency risk is significantly negative, at −0.44. Even though the currency volatility is quite high at 10.72%, the volatility of unhedged equity returns is lower than the hedged returns. In other words, the foreign currency exposure provides diversification to equity risk for Australian investors whereas in the case of Japanese investors currency exposure adds to equity risk. Therefore, from a risk management perspective, currency risk could provide a hedge to equity risk for Australian investors. However, this hedging is not cost free. Currency exposures in the equity index had a negative return of 3.38%, which is even worse than the negative 2.69% for the bond return index shown in Table 2.19.

2.3.5 Carry Risk Premium

Diligent readers might have sensed that while there is no broad-based currency risk premium, like equity or interest rate, there might be selective currency risk premium with respect to certain base currencies. For example, currency risk seems to bring positive returns to JPY-based investors while it brings negative return to AUD-based investors.

Of course, this is not a random result. One of the most popular trades in currency is the carry trade, in which one invests in high-yielding currencies at the expense of low-yielding currencies. Over the last two decades, as we alluded above, among all developed countries Japan has had the lowest short-term interest rate due to prolonged deflation, while Australia has had the highest short-term interest rate partly due to its status as a natural resource producer. Supposedly, Japanese housewives have been on this trade for a long time by putting money into AUD-denominated assets. On the other hand, one never hears about Australian housewives doing the same by piling into JPY-denominated assets.

On average, long AUD and short JPY crosses, or carry trades in general have been profitable. This is why JPY-based investors have had positive returns from foreign currency exposures whereas AUD-based investors have been on the losing side of currency exposures.

Nevertheless, it is easy to see currency carry premium is not like equity risk premium or interest rate premium. It is better understood as a factor risk premium that is similar to a value premium. In the case of carry, short-term interest differential is similar to other value metrics like dividend yield in equities or yield curve slope in fixed income. The other similarity is it requires both long and short positions to harvest the carry factor premium.

Carry trade strategies are highly correlated with equity market returns, however. In Figure 2.4, we plot the cumulative returns of the MSCI world USD-hedged index return and the G10 carry trade index constructed by JP Morgan. For most of the time since 1995, the two lines move together, except in years from 2000 to 2002 where the equity market declined in the recession after the dotcom bubble burst. Starting in 2003, the two have been in almost perfect alignment. The correlation of the two monthly return series for the entire period is 0.52.

For both Japanese and Australian investors, the return from foreign currency exposure mirrors the return from carry trades. Figure 2.5 plots again the cumulative return of carry,

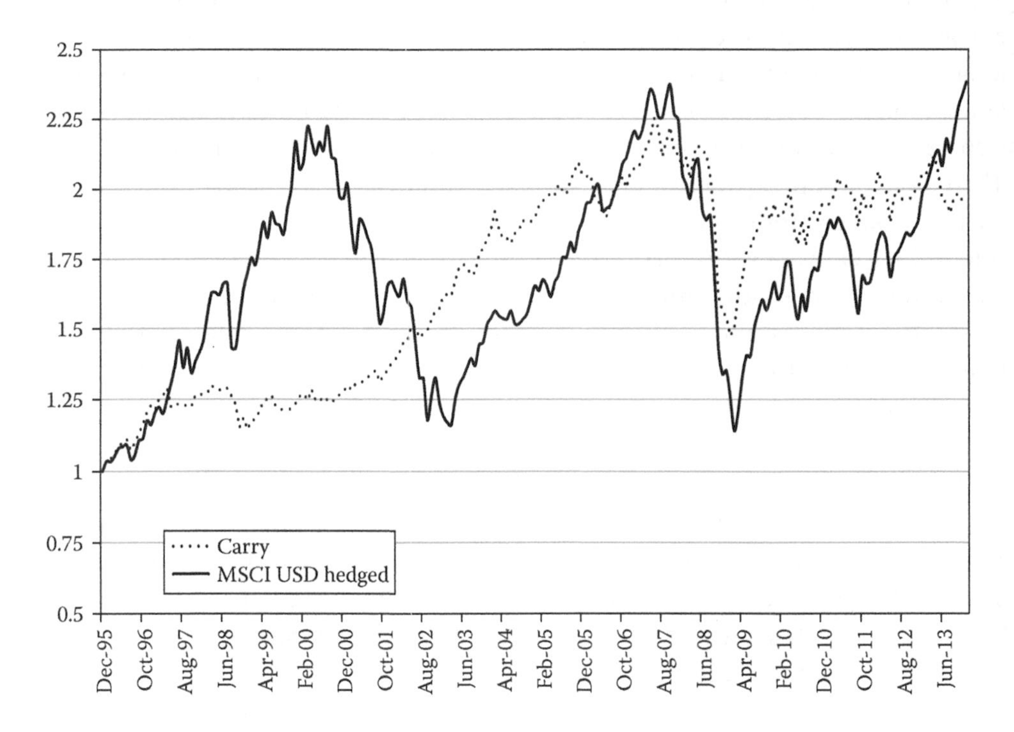

FIGURE 2.4 Cumulative returns of MSCI World Index (USD hedged) and carry trade.

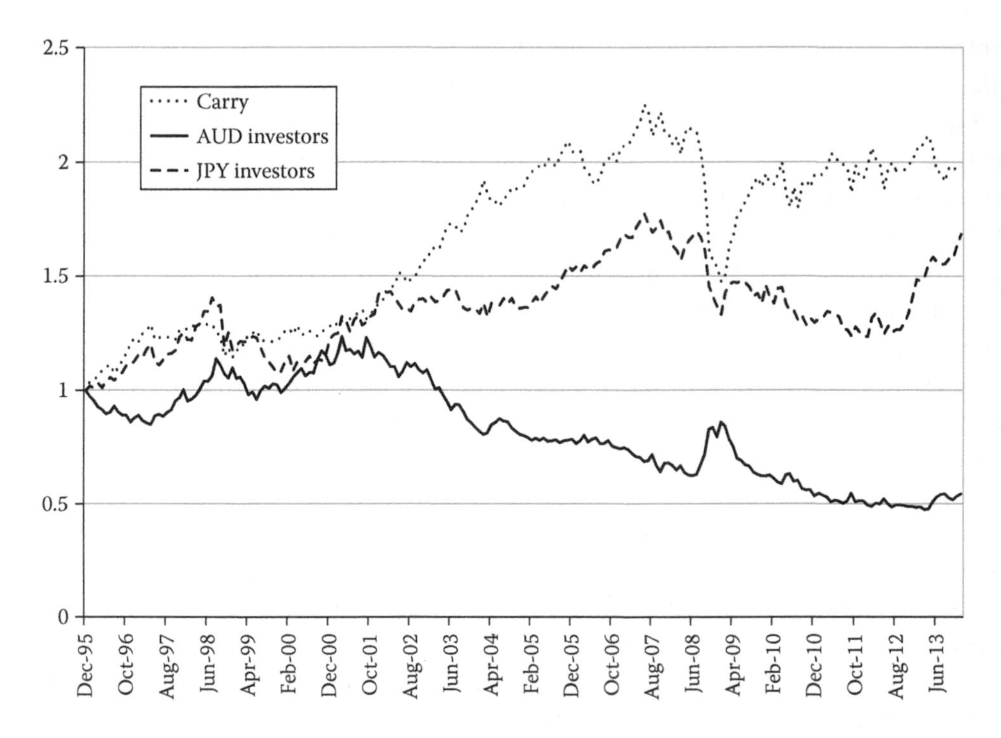

FIGURE 2.5 Cumulative returns of carry trade, and currency returns from unhedged MSCI world indices for Japanese and Australian investors.

together with the return of foreign currency exposure from both a JPY and AUD perspective. For Yen investors, foreign currency exposure in the unhedged equity index is a form of long carry. The correlation between the two is 0.57. On the other hand, for Australian investors, foreign currency exposure in the unhedged equity index is a form of anticarry. The correlation between the currency return and the return from carry is −0.73!

Therefore, for both Japanese and Australian investors, the question of whether to invest in currency-hedged indices or unhedged indices are akin to the question of whether they should be long carry or short carry. This decision of how to treat carry premium should be made jointly with one's allocation to traditional market risk premiums such as equity and interest rate premiums, and to other investment strategies within a total portfolio. In addition, it is important to note that other risk premiums, such as interest rate risk premium, might offer a more cost-effective hedging for the equity returns.

2.3.6 Conclusion

Risk parity asset allocation portfolios invest in market risk premiums with a balanced risk allocation. These market risk premiums include equity, interest rate, and inflation risk premiums. Not all risks belong in a risk parity portfolio. For example, it is not desirable to have illiquid assets such as HY bonds since risk parity portfolios with leverage requires regular portfolio rebalancing.

In this research note, we discuss the role of currency risk. We have shown that, from both a fundamental and an empirical perspective, currency risk does not have any intrinsic risk premium. In other words, foreign exchange exposure does not bring excess return relative to the domestic risk-free rate. Therefore, investing in equity and interest rate risk premiums in developed foreign markets should be done on a currency-hedged basis.

For emerging markets (EM), fully hedging currency exposure is impossible for a variety of reasons. Even there, we have found that there is no empirical evidence of currency returns when we compare the MSCI EM index in local terms versus the MSCI EM index return in USD terms.

One form of "currency risk premium" often referred to in academic research is the return from the carry trade. Carry trades are implemented through long–short portfolios that buy high-yielding currencies and sell low-yielding currencies. It is better described as a factor risk premium such as a value premium or momentum premium. The sources and durability of such factor risk premiums are not the same as traditional market risk premiums. Nevertheless, they can provide additional diversifying returns beyond market risk premiums in conjunction with a thoughtful portfolio construction process (Qian et al., 2013).

Our results do raise an interesting question concerning the combination of market risk premiums and factor risk premiums for investors with different base currencies. This is especially true for carry factor premium, because of its positive correlation with equity market returns and its reliance on long and short positions in different currencies. For investors in low-yielding currencies, it might be desirable to be long carry to increase total returns through carry. Conversely, for investors in high-yielding currencies, it might be desirable to be short carry to reduce overall portfolio volatility at the expense of carry. However, focusing solely on currency carry to diversify equity risk is not enough. In all cases, a careful analysis of risk/return tradeoff might offer a way to optimally combine market risk premiums and factor risk premiums together for different investors with different currency perspectives.

CHAPTER 3

The "Death" of Interest Rate Risk Premium

By far, the most frequently asked questions about risk parity are questions about interest rates. A few examples are: What happens to risk parity when interest rates go up? What happens when the Fed raises interest rates? What happened in 1994? Personally, I have never been in a discussion with prospects or investors regarding risk parity in which one of these questions or some variation of it was not asked.

This singular concern on interest rates is both understandable and rather puzzling. As we discussed in Chapter 1, interest rate risk is only one of the three (sometimes four) risk premiums in a risk parity portfolio. Even though the notional exposure (weight) to bonds is high for the sake of balancing its risk contribution with that of high volatility assets, its potential impact on portfolio return is no greater than that of stocks. As we shall demonstrate in Chapter 4, the diversification benefit of risk parity, in general, can offset potential losses suffered by interest rate exposure in a properly designed risk parity portfolio. So, why focus so much on bonds but not on stocks or commodities? One potential explanation is that some might view investing in risk parity as an active decision of buying bonds and selling stocks in lieu of a traditional 60/40 portfolio and are worried about the timing of this action. However, a deeper reason seems to come from an either explicit or implicit forecast that interest rates will surely rise. If rates were to rise persistently, then there would be no return premium or worse negative return for bearing interest rate risk, or the "death" of interest rate premium.

There are two issues for this dire view of interest rate premium. First, forecasting future interest rates is not as easy as it seems. In fact, forecasting anything in the future is hard. However, this has not prevented experts, strategist, and pundits to make forecasts of rising interest rates for the last five years after the global financial crisis. Unfortunately, there has not been lack of materials for making such a forecast. The first is a pure technical forecast. After years of declining interest rates caused by declining inflation, it is rather tempting to make a forecast of mean-reversion in the way of higher interest rates in the future. The second reason for arguing for higher interest rates was fundamental in nature. Ever since the Fed and other central banks embarked on quantitative easing (QE)* in order to spur economic growth after short-term interest rates of many countries hit the zero bound, the conventional wisdom was QE would lead to massively higher inflation and higher interest rates.

Other forecasts of higher interest rates are more analytical and nuanced. For instance, when short-term rates are near zero, the yield curve is always upward sloping since long-term yields usually stay positive. An upward sloping yield curve leads to forward interest rates that are higher than current rates. Some took this as a sign of higher future rates.

Yet another popular argument for reducing interest rate exposure stemming from the perspective of risk control is to equate the duration of fixed-income assets to risk. As yields decline, durations of fixed-income assets extend further. If duration equates to risk then one must reduce fixed-income exposures.

All these arguments are either wrong or not even wrong. This is the case, not because the prediction of higher interest rates turned out wrong. Interest rates have actually declined instead of rising after these arguments were made. This is the case because the arguments for higher interest rates and for reducing interest rate exposure are not logic in the first place. In this chapter, we present three investment insights on the subject of interest rates. The first made the point that bond yields had been low partly for fundamental reasons. Hence, mean-reversion of bond yields may not happen unless the macroeconomic environment changes. The second insight dispels the notion of using duration as the risk measure for bond returns. The third insight makes the point that forward interest rates are not a reliable forecast of future interest rates.

It is worth noting that none of the arguments presented here is forecasting lower interest rates in the future. The emphasis of the arguments is about the difficulty and potential bias of making forecasts on future interest rates. If possible, one should resist such urge no matter how easy it might seem.

3.1 ARE BOND YIELDS TOO LOW?†

Ten-year US treasury (UST) bond yield, probably the most noted barometer of long-term interest rates in the United States is near 3% as of August 2011. It has been stuck in a range from 2.5% to 3.8% since June 2009, when the last recession ended and recovery started, according to the National Bureau of Economic Research (NBER). For quite some time now, there seemed to exist a universal consensus among the majority of sell-side forecasters, buy-side investors, and TV pundits that, bond yields would soon go up, and probably go up significantly. The list of reasons is quite long but changing, from strong economic growth, to high commodity prices, to rising deficits, to QE2, to US sovereign credit risk.

However, the treasury market has so far refused to validate this view. The question remains though: Are bond yields too low? As we demonstrate in this note, the answer is not as straightforward as some might think. A related question is should one invest in

* Quantitative easing is otherwise known as asset purchasing program, in which central banks buy government bonds and other fixed income assets from banks and private investors.

† Originally written by the author in August 2011.

bonds at this yield level? The answer to this second question is not necessarily the same as the answer to the first.

3.1.1 The Answer Is Yes

One of the most commonly quoted reasons for higher yields is actually not among the list above; it is far simpler: the current yield is just too low according to historical standard. Figure 3.1 shows the 10-year UST yield since 1980, which has declined steadily from above 10% to 3% today. With zero being the floor for nominal bond yields, it is rather hard for many to admit the possibility that the yield can go lower. On the other hand, it is far easier to envision the yield will bounce up. Mean reverting does have its appeal.

Of course, this argument is too simplistic. Even in cases where mean reverting does eventually occur, timing is always of the essence.

Thinking back several years, the same argument could have been made in 2002, and again in 2006, when the yield was quite low compared to historical levels—the former Fed Chairman Greenspan called it a conundrum (Federal Reserve Board's Semiannual Monetary Policy, 2005). Many investors presumably abandoned government bonds in favor of bonds with higher yields and risky assets. Those predictions proved premature and probably costly as well.

Another counterexample against the simplistic argument comes from the Japan experience. Figure 3.2 shows the 10-year Japanese government bond (JGB) yield since 1984. Imagine, back to 1995 when the 10-year JGB yield was near 3% and pretend that we could not see the line inside the shaded region. Could one reasonably claim that 3% is too low because it is near the historical low? The subsequent event proved that it was not to be the case. Today the 10-year JGB yield hovers a little above 1% and few people claim it is too low. Maybe it is. Indeed, it is hard to resist making such a claim when you just stare at the chart.

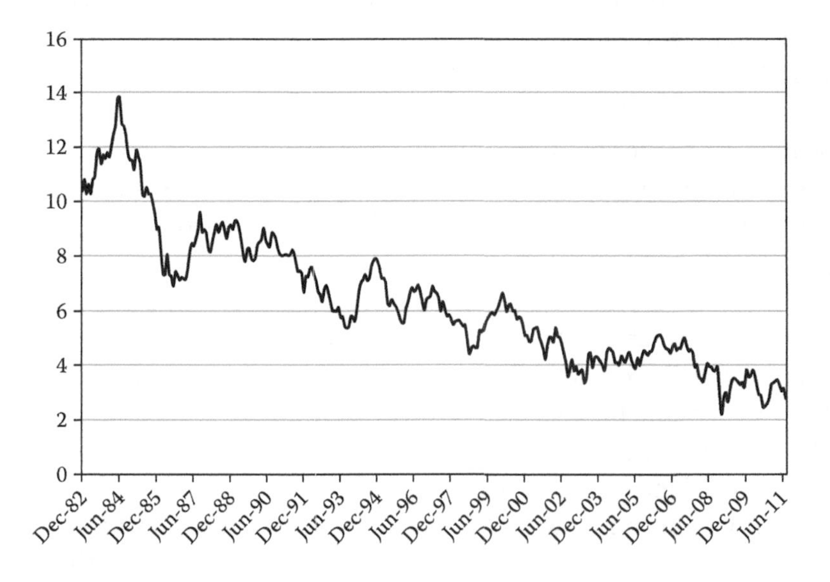

FIGURE 3.1 UST 10-year yield.

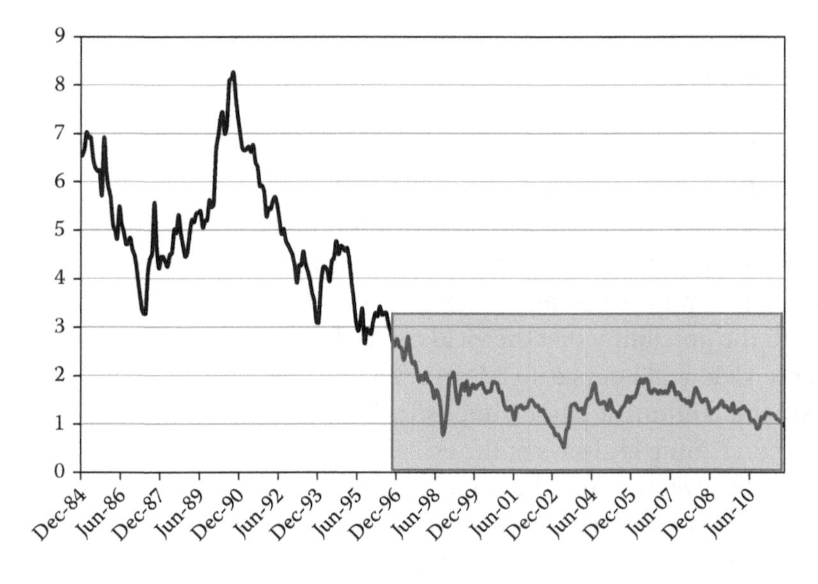

FIGURE 3.2 JGB 10-year yield.

3.1.2 The Answer Is Maybe Not

The reason that the naïve view of reversal could be wrong is it ignores the economic fundamentals of interest rates. Theories of interest rates are too numerous to repeat here. Empirically, there are at least two important economic factors related to long-term interest rates. The first is the long-term inflation of underlying economy. This is intuitive because investors in long-term government bonds need to be compensated for purchase power erosion due to higher inflation. The second factor is real growth. The intuition behind this link is twofold. One, maybe the real growth is an opportunity cost for bond investors who forsake investments in the real economy. Two, real growth is a reasonable proxy for term premium between long-term bonds and short-term cash.

One might then conclude that nominal growth that contains both growth and inflation components would suffice to determine the long-term interest rate. This is true—nominal growth is a good anchor for the 10-year bond yield. However, our research has found that using the two components separately and replacing headline consumer price index (CPI) with core CPI, which is much more stable and a better proxy of long-term inflation, lead to higher explanatory power.

Figure 3.3 plots the 10-year UST yield and fitted value based on a linear model with the two aforementioned variables. On a contemporaneous basis, the fit is very strong in terms of both secular trend and cyclical changes. For the most recent quarter Q2 2011, the core CPI is 1.5%. If we assume the real GDP growth for Q2 2011 is 1.5%, then the fitted yield would be 3.28%, versus the actual yield of 3.16%. Therefore, at least based on in-sample data, the 10-year yield may not be too low.

Of course, if inflation and growth pick up in the future, one would expect the yield to go up. But in the absence of heightened inflation risk and above-trend growth, it is hard to see long-term interest rates go up significantly.

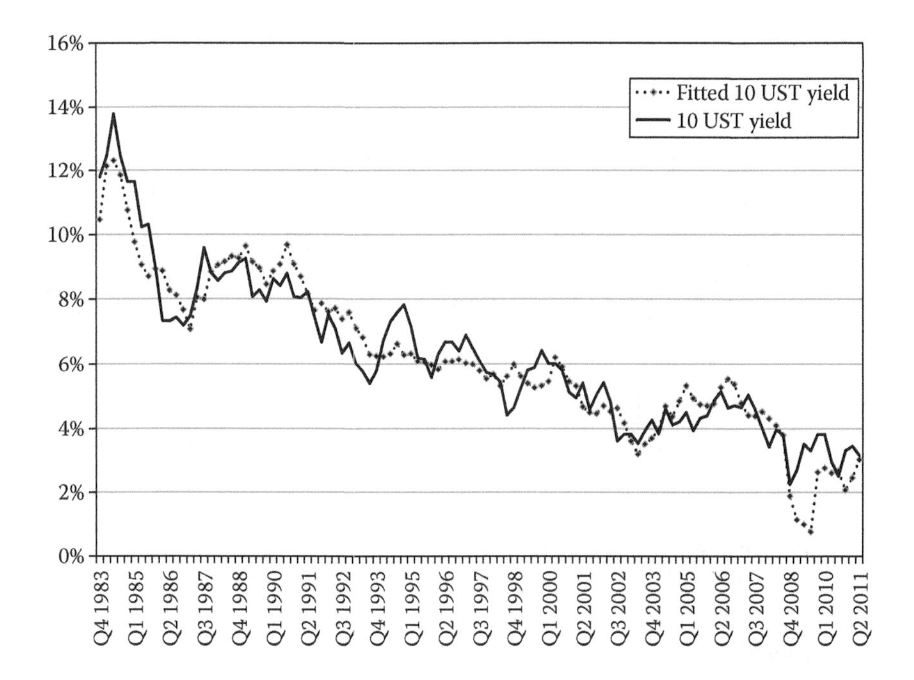

FIGURE 3.3 Fitted and actual UST 10-year yield.

3.1.3 The Answer Is Actually No

So far, we have attempted to make the case that even though the current yield appears low relative to its historical level, it may not be so relative to the underlying economic fundamentals. Investments and valuations are all relative. Indeed, relative to another investment, not only is the current yield not low, it actually looks attractive. That investment is cash, which is currently yielding close to zero. In other words, the spread between the 10-year yield and the cash rate—one version of yield curve slope, is rather steep, even by historical standards.

Figure 3.4 shows the yield curve slope since 1982, which mostly varies between 0% and 4%. On a few occasions when it inverted, dipping below 0%, recessions ensued. When the US economy subsequently recovered, the yield curve steepened, pricing in an expected rise in growth and inflation. Over this period, the maximum steepness is close to 4% and the average has been about 2%.

With the current cash rate being close to zero, the slope of the yield curve is roughly 3%, making it higher than the historical average. Thus, the implication of a steep yield curve is that the expected excess return of bonds over cash is potentially attractive.

We can view this attractiveness in two ways. First, if both the long-term bond yield and the short-term interest rate stay where they are for the next year, then the excess return of a 10-year bond will be roughly 3%. A 3% return might not be something to write home about, but relative to the risk level of bond returns, say 6%, its risk-adjusted return, that is, Sharpe ratio, is 0.5. A Sharpe ratio of 0.5 is attractive. To put it in perspective, for US large cap equity market to achieve the same Sharpe ratio, the return needs to be at 8% since the risk is at least 16%.

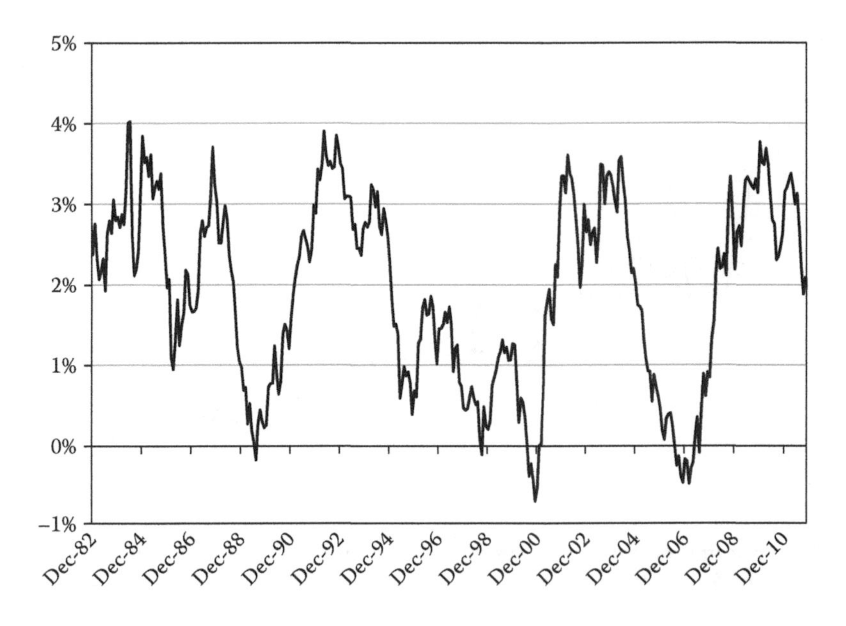

FIGURE 3.4 UST yield curve slope.

Second, our research shows that yield curve slope has predictive power for future excess returns in bonds. Figure 3.5 plots the yield curve slope and the 1-year forward excess return of the Citigroup UST Bond Index. The two lines show significant co-movement and their correlation is near 0.3. While not a perfect indicator, steeper yield curve, like the one we have today, does indicate higher excess returns for bonds.

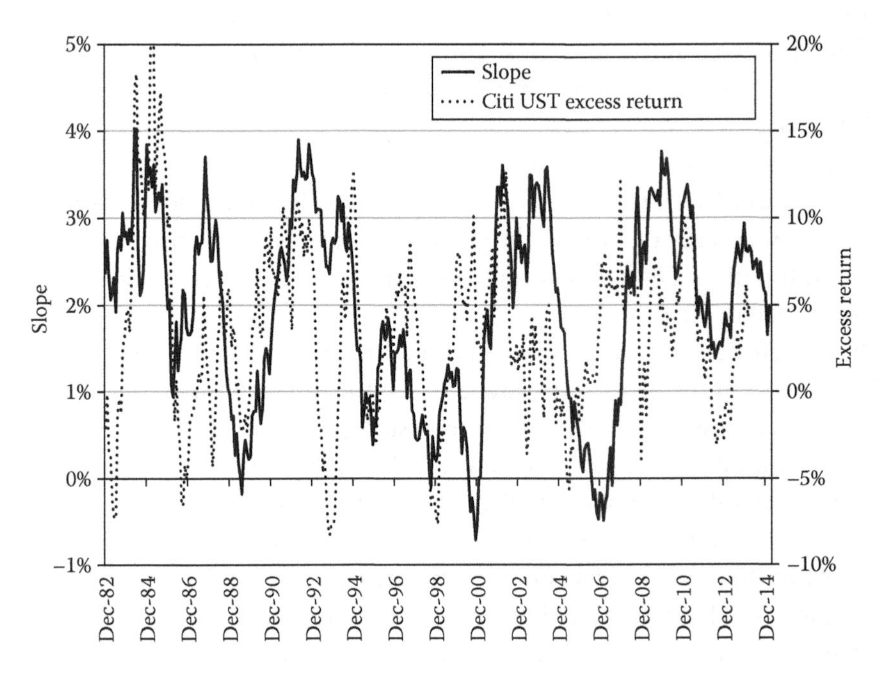

FIGURE 3.5 Yield curve slope and 1-year forward excess return.

3.1.4 The Answer Is it Depends

Forecasting future asset class returns is never easy. This is equally true for both stocks and bonds, even though the latter seems to invite the illusion that easy answers are attainable, since only one factor—yield, is all one needs to consider. Investors should try to avoid such illusion because that single factor is an aggregation of a multitude of underlying factors, including economic fundamentals and yield curve dynamics.

In summary, although the current treasury yield might appear low relative to the historical standards, it may not be so relative to the economic fundamentals in the United States. Furthermore, the bond yields, those of 2-year, 5-year, and 10-year treasuries actually appear quite attractive in terms of return premium over cash, making them suitable investments in multi-asset portfolios, especially so for risk parity.

3.2 DURATION, YIELD VOLATILITY, AND BOND EXPOSURE*

Duration is probably the most important risk measure used in managing fixed-income assets. It measures the sensitivity of a bond's price return to changes in interest rates. For example, a bond with a 1-year duration would deliver a price return of approximately +1% or −1% if interest rates instantaneously declined or rose by 1%, respectively. Everything else equal, a higher duration means a higher level of risk for a bond.

Another fact of fixed-income mathematics is durations of coupon-bearing bonds lengthen as interest rates go lower. This makes sense, since the overall duration could be thought of as the weighted average of durations of all cash flows of a bond. Consequently, lower rates lead to lower weights for shorter-duration interest payments and a higher weight for the long-duration principal payment and as a result longer duration.

For the last 4 years since the 2008 global financial crisis, a weak global economy with low levels of inflation have contributed to a trend of declining sovereign interest rates to historically low levels. As interest rates have declined, bond durations have extended. Therefore, from a duration perspective, all these government bonds appear riskier than ever. Simply focusing on duration would lead one to believe that portfolios with fixed-income positions such as risk parity would need to reduce its bond exposure in order to maintain its targeted risk allocation.

Despite its intuitive appeal, this is a faulty conclusion because beside duration, everything else is not equal. Even though the durations of these sovereign bonds have lengthened, yield volatilities, in many cases, have also collapsed, partly due to market cycles and partly due to unconventional global monetary policies. Since volatility of price return is a product of duration and yield volatility, the change in the riskiness of a bond's price return depends on the relative magnitude of changes in the two components. In this research note, I shall examine the recent changes in yield volatilities for some major sovereign bond markets. In many cases, the declining yield volatility dominates the prolongation of duration and as a result, the return volatilities of these sovereign bonds have actually declined rather than risen. I will then discuss how this phenomenon impacts the sovereign bond exposures in a risk parity portfolio.

* Originally written by the author in December 2012.

TABLE 3.1 Duration of a 10-Year Bond with Different Yields

Interest rate	5%	4%	3%	2%	1%
Duration	8.0	8.3	8.7	9.1	9.5

3.2.1 Duration with Declining Interest Rates

The duration extension due to declining interest rates is a gradual process. To see this explicitly, we list the duration of a 10-year bond with interest rates ranging from 1% to 5%. To simplify the analysis, we assume these are par coupon bonds (i.e., yield is the same as the coupon rate). At 5%, the duration is about 8 years and it increases to 9.5 years when the interest rate declines to 1%. If the interest rate drops to zero, the bond becomes a zero-coupon bond and the duration equals its maturity, that is, 10 years. In other words, there is not a lot of room for duration to increase as the rate goes lower (Table 3.1).

Over the course of the last 5 years, the US 10-year treasury yield has dropped from around 4% at the end of 2007 to 1.7% at the end of 2012, while its duration has increased about 11% from 8.3 years to approximately 9.2 years.

The sensitivity of a bond's duration to changes in interest rates is greater for longer-maturity bonds. Table 3.2 lists the duration of a 30-year bond with the same interest rate scenarios. Notice that the difference in duration between 5% and 1% is close to 10 years while it is only 1.5 years for the 10-year bond. On the other hand, for a 5-year bond or a 2-year bond, the duration barely moves as the interest rate varies. This is because for shorter maturity bonds, the duration is mostly determined by the time horizon of the return of the principal rather than those of the coupon payments. In other words, "for shorter maturity bonds return volatility is mostly determined by yield volatility."

3.2.2 UST Yields and Yield Volatilities

A confluence of factors has led to the fall in UST yields over last few years: anemic economic growth, low inflation due to deleveraging, multiple rounds of QE, and more and more explicit forward rate guidance by the Fed. Across the term structure, while the yield curve as a whole is interconnected, the long end of the yield has more to do with macroeconomic variables such as growth and inflation whereas the short end has more to do with the Fed's interest rate guidance.

Figure 3.6 displays the yields of 2-, 5-, 10-, and 30-year UST bonds since 1995. While all four yields have trended downward, yields on the front end of the curve as represented by the 2- and 5-year yields have become especially depressed. They are substantially lower than the levels breached during the height of the global financial crisis in 2008. In contrast, the 10-year yield now is only 50 basis points lower than at the end of 2008 and the 30-year yield today is even a touch higher than at the end of 2008. This steepening of the yield curve has had a meaningful impact on changes in yield volatility across the yield curve.

TABLE 3.2 Duration of a 30-Year Bond with Different Interest Rates and Yield

Interest rate	5%	4%	3%	2%	1%
Duration	15.8	17.7	20.0	22.7	26.0

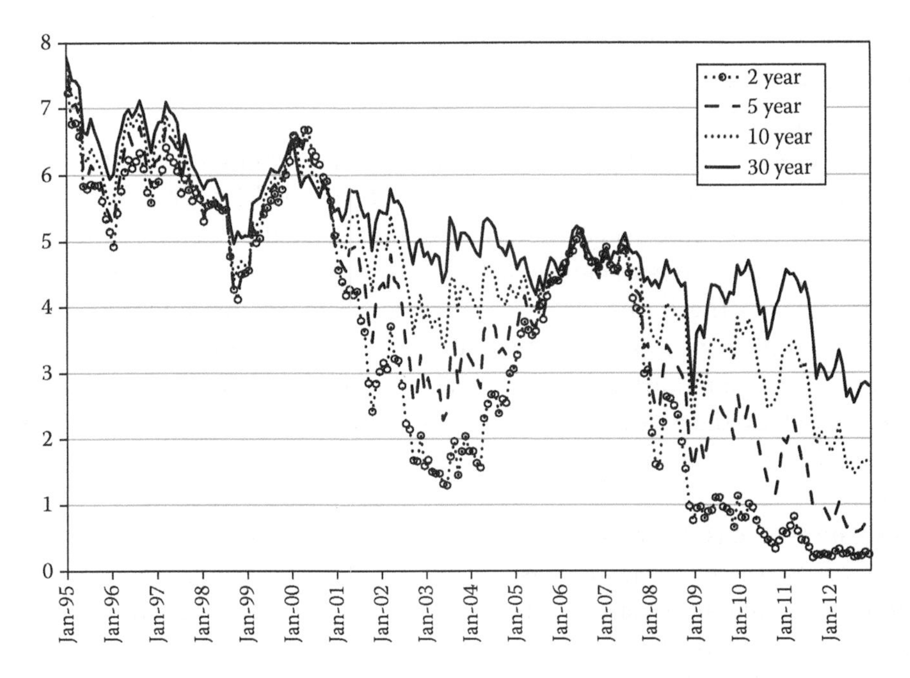

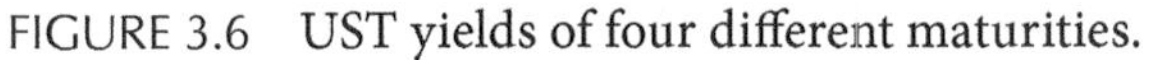
FIGURE 3.6 UST yields of four different maturities.

For yield volatility, we calculate the 12-month standard deviation of monthly changes in yields and then annualize it by multiplying the same by the square root of 12. Figure 3.7 displays the four yield volatilities.

There are several important features in this graph. First, yield volatility varies greatly over time. Even though this might be influenced by the rolling window of 12 months in

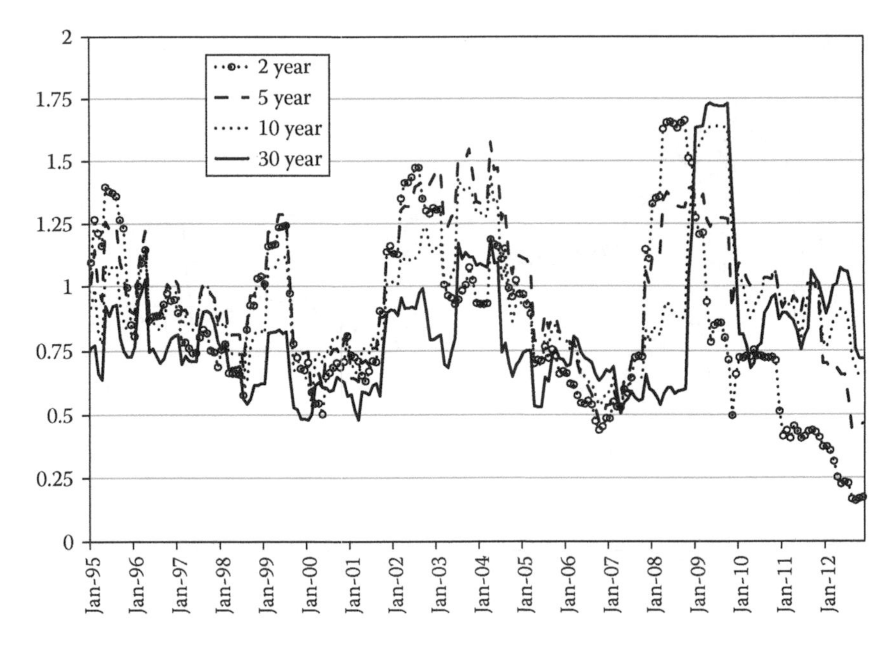

FIGURE 3.7 Annualized standard deviation of rolling 12-month change in yields.

our calculation, the variability persists with longer time windows. In general, the yield volatility trends downward during economic expansions and spikes up during recessions.

Second, prior to the most recent period, the yield volatility is higher for short-term yields and lower for long-term yields. In normal interest rate environments when the Fed Funds rate is far away from the zero bound, the 2-year and 5-year yields are the most sensitive to macroeconomic expectations and the anticipated monetary policy path of the Fed. However, this order has been reversed since 2011. Now the 2-year yield has the lowest volatility, followed by the 5-, the 10-, and the 30-year yield. The reason for this reversal lies in the Fed's commitment to zero interest rate policy (ZIRP)—for the foreseeable future. This commitment first rendered the 2-year yield irresponsive to any macroeconomic news and it has now made the 5-year yield more or less static. In contrast, the 10- and 30-year yields are still reacting to macroeconomic surprises, exhibiting a high degree of yield volatility.

To evaluate the sensitivity of yield changes to changes in the economy we estimate yield betas across the curve for changes in the Institute for Supply Management (ISM) Manufacturing Purchasing Managers Index (PMI). The yield betas capture the sensitivity of monthly changes in yield to changes in the US PMI.

Figure 3.8 plots the yield betas relative to their historical averages since 1993. As you can see from the chart, both the 30- and the 10-year parts of the curve currently exhibit above-average sensitivity to changes in the PMI, while the 5- and 2-year points of the curve show below-average sensitivity to changes in the PMI index.

Third, in terms of absolute level, both volatilities of the 2- and 5-year yields have broken out of their historical ranges and sunk to new lows while the volatilities of the 10- and 30-year yields are in the middle of their historical ranges (Figure 3.7). The implication is

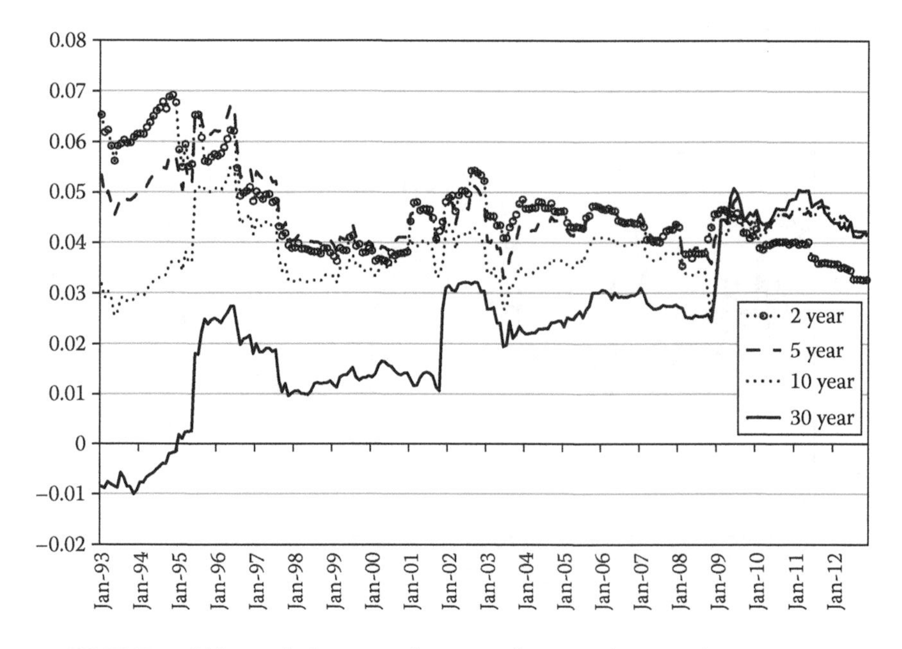

FIGURE 3.8 US PMI yield betas (relative to their own historical average).

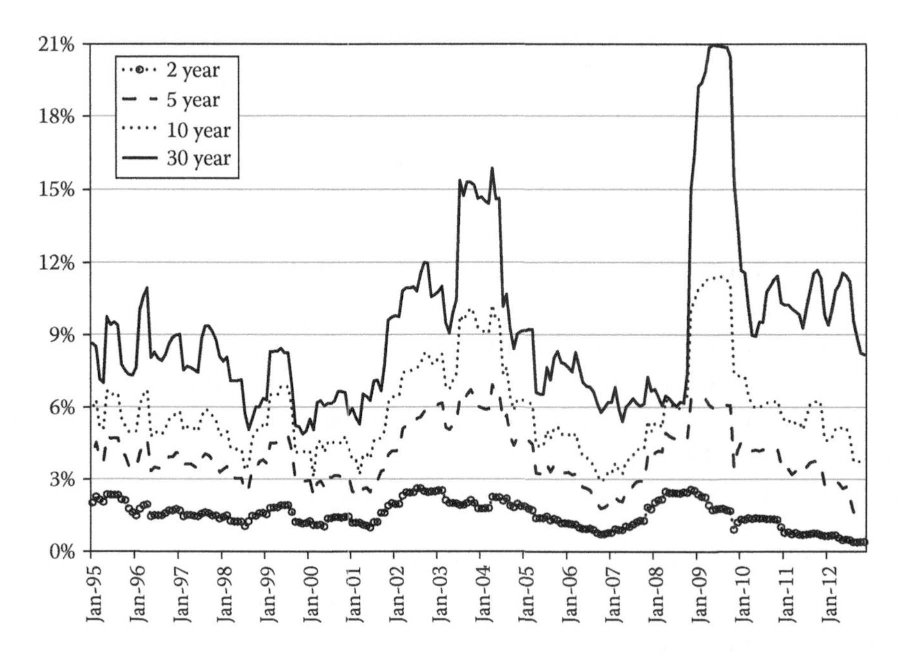

FIGURE 3.9 Annualized standard deviation of monthly UST future returns with a rolling 12-month window.

that while 2- and 5-year bonds are much less volatile than they have been historically due to a collapse in their yield volatility, 10- and 30-year bonds still have plenty of price return volatility because of their extended duration and their average yield volatility.

The last observation proves that the return volatilities of UST bonds would have declined in the recent period despite their higher durations. Indeed, this is confirmed by Figure 3.9 that shows the annualized return volatilities of four UST futures. Note that similar to the yield volatility, the 2-year note future's return volatility has sunk to a new low. The 5-year note future's volatility is similarly at a historically low level. However, for the 10-year note future and 30-year bond future, the return volatilities are within their historical ranges even though they have been trending lower since 2009.

3.2.3 Risk Parity Treasury Exposures

What does all this mean for risk parity in terms of its exposure to UST bonds? First, it is beneficial to build exposure to USTs with a risk parity approach across different maturities to diversify interest rate risk across the term structure. The benefit of this approach versus a single spot exposure on the yield curve—say the 10-year note future, is that risk parity across the term structure balances the risk allocation across short- and long-term interest rate risk thus providing better risk-adjusted returns. In addition, one can also dynamically shift risk allocation along the segments of the yield curve if one is able to incorporate active views of future yield curve movement.

A dynamic process can play an important role in mitigating today's environment of low yields and low-yield volatilities. Without it, the notional exposure to the front end of the curve (2-year and 5-year), would increase substantially, due to the historically low-return volatilities described above. While increased leverage to the short end of the yield curve that arguably offers little term premiums might provide hedging against a steepening curve, it nevertheless represents an inefficient use of a portfolio's risk budget. In a way, in the current environment, "the 2-year bond is the new 'cash' and the 5-year bond is a new '3-year bond.'"

In contrast, the long end of the yield curve offers relatively attractive term premium. In addition, their return volatilities have not declined as sharply as short-term yields and are currently near their historical averages. Therefore, we need not raise the leverage by much in order to maintain our targeted risk allocation to UST bonds as a whole. Furthermore, from a portfolio perspective, since both the 2-year and to some extent 5-yield yields are anchored by ZIRP, they no long provide useful hedging against the equity risk exposures of a risk parity portfolio. In contrast, the long end of the yield curve still exhibits negative correlation to equity. Therefore, an increased risk allocation to 10- and 30-year futures is appropriate from a portfolio diversification perspective.

3.2.4 Global Trend in Yields and Yield Volatilities

In addition to the US experience, other high-quality sovereign bonds have exhibited similar changes in yields and yield volatilities, with some notable exceptions. The synchronization in the movement of global yields is not all that surprising. First, the global recovery has been weak across many developed countries. Second, central banks around the world had cut policy rates aggressively to very low levels. Third, some central banks, such as the Bank of England, Bank of Japan, even European Central Banks, have joined the Fed in implementing QE with the goal of lowering interest rates.

Figure 3.10 shows the 2-year sovereign yields for six countries, all of which are at or near historically low levels. The JGB 2-year yield has been at a very low level for decades as the

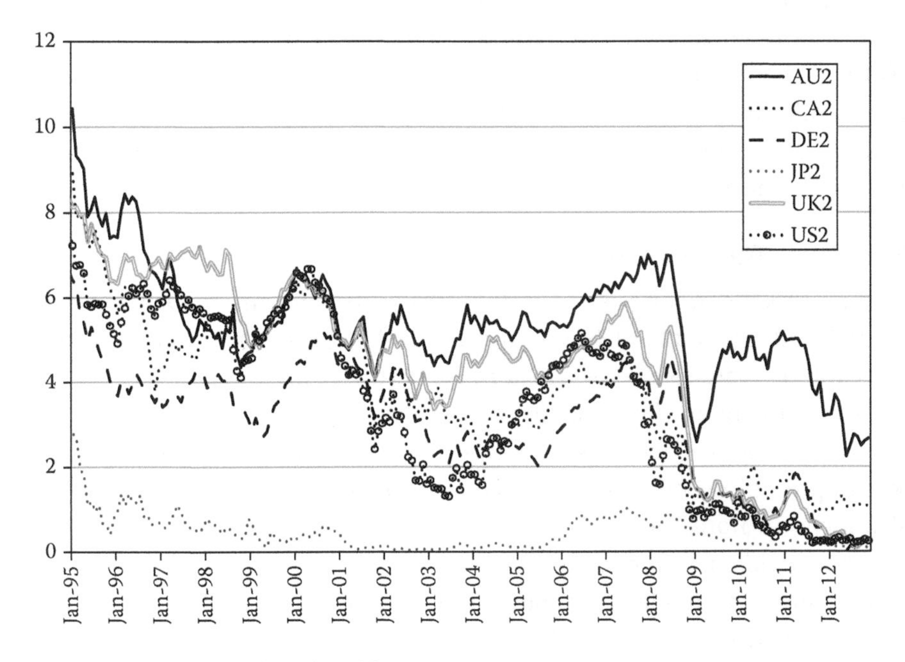

FIGURE 3.10 2-year sovereign bond yields.

country has been mired in persistent deflation. Among the six countries, only Australia and Canada, whose economies have a large commodity component, still have high 2-year yields, in relative terms, of course.

Similar to Figure 3.9, Figure 3.11 displays the yield volatilities for the 2-year yields of the six countries. Again, we note that all have fallen either below or near historical lows, except for the Australian 2-year yield. Other countries have since joined Japan in their races to the bottom, in terms of both the level of yields and the level of yield volatilities. Only in Australia, whose policy rate is still far above zero, is the short-term yield still responsive to macroeconomic variables such as commodity prices and economic conditions in China.

The situation in 10-year yields mirrors that of 2-year yields. As we can see in Figure 3.12, while all yields have dropped significantly over the recent years, Japan and Australia are two "outliners," with Japan's 10-year yield below 1% and Australia's 10-year yield still close to 3%. A case can be made that the former offers little risk premium while the latter looks more attractive from a valuation perspective.

With the lowest yield among its peers, JGBs have always had the longest duration in the group. However, duration comparison across different countries is meaningless in terms of assessing the riskiness of sovereign bonds. The reason is yield volatilities are surely different across different countries. As it turns out, the 10-year JGB has markedly lower yield volatility compared to other 10-year sovereign bonds.

This is evident in Figure 3.13, which shows the 10-year yield volatilities for the six countries. Several features are worth noting. First, similar to the United States, with the exception of Japan, 10-year yield volatilities of all other countries have declined recently but they still lie within their historical range. Second, the 10-year JGB yield volatility is as low as that of the 2-year UST yield (see Figure 3.11). In many ways, the 10-year JGB is similar

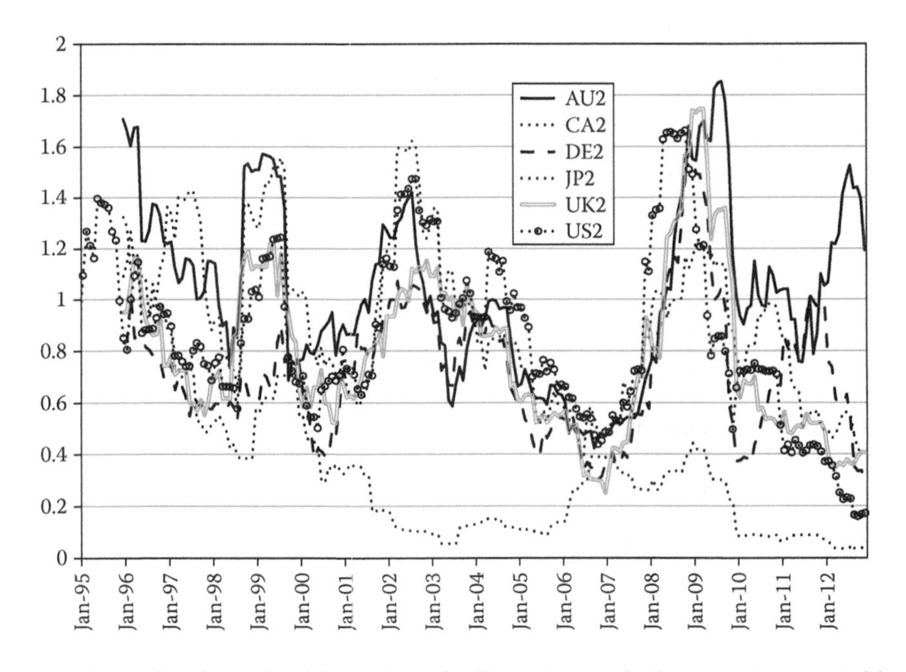

FIGURE 3.11 Annualized standard deviation of rolling 12-month change in 2-year yields.

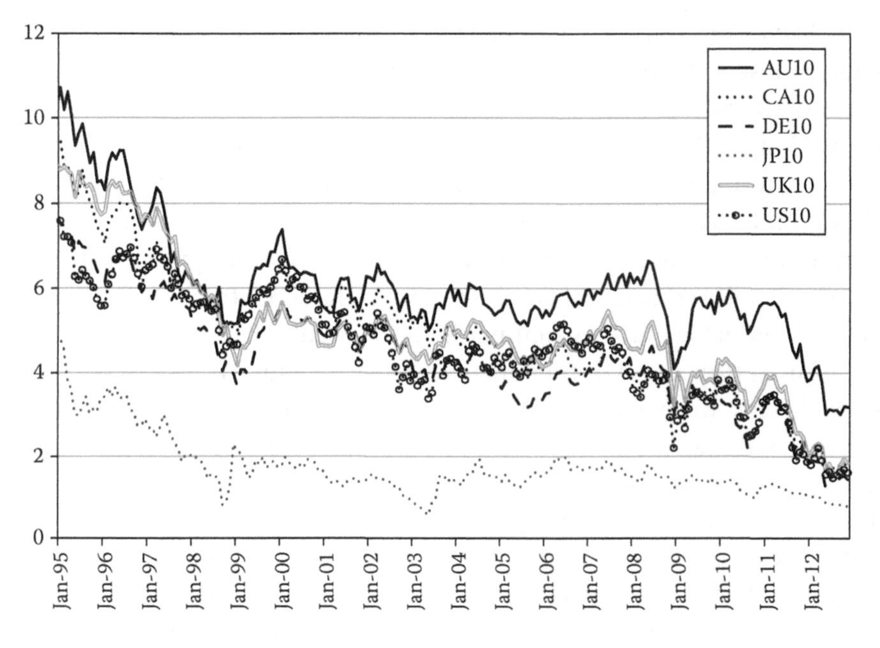

FIGURE 3.12 Ten-year sovereign bond yields.

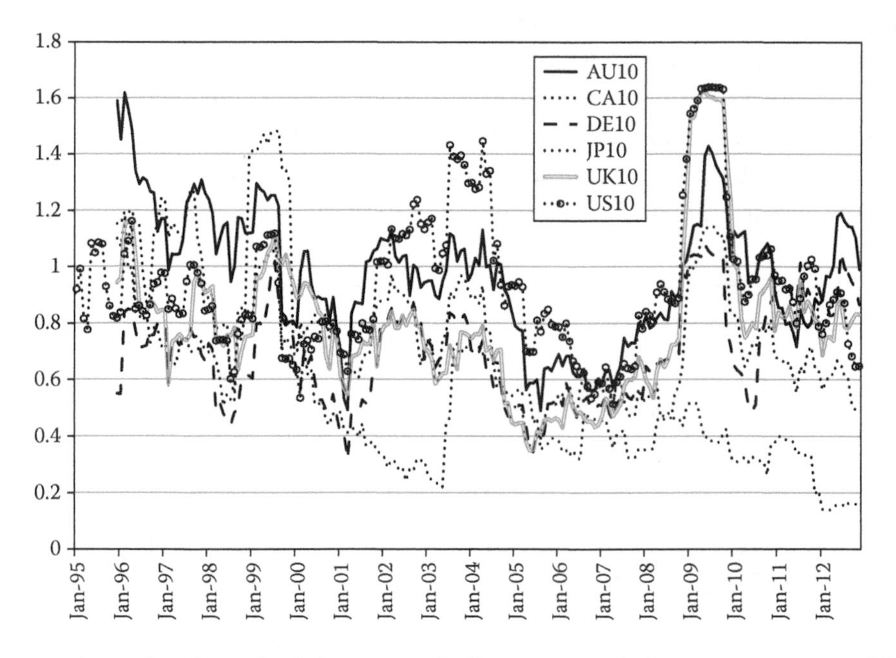

FIGURE 3.13 Annualized standard deviation of rolling 12-month change in 10-year yields.

to a 2-year UST bond, in terms of its low yield and low-yield volatility. Similarly, it can be argued that it offers little risk premium.

3.2.5 Conclusion

Risk measures are an integral part of risk parity portfolios. For fixed-income exposures, it might seem that duration serves as a useful risk measure. However, as we have shown above, duration by itself does not adequately capture the riskiness of fixed-income assets across different maturities and different countries. Rather, yield volatility, in many cases, is more important in determining the riskiness of fixed-income assets. The key is one should not use duration alone as the risk measure of a bond.

In the current market environment, the yield volatilities of 2- and 5-year UST bonds are exceptionally low, leading to a lower risk estimate for both. Instead of raising their notional exposure to these assets to maintain its targeted risk allocation, one might consider reducing their risk allocations while increasing the risk allocation to 10- and 30-year bonds, which seem to offer better risk premiums and a return volatility close to its historical average.

Across other high-quality global sovereign bonds, the same phenomenon manifests itself in extremely low yields and low-yield volatility in JGBs. A similar solution supported by fundamental intuition would be to reduce the risk allocation to JGBs and increase the risk allocation to other countries with better interest rate risk premiums.

3.3 DO FORWARD RATES HAVE ANYTHING TO DO WITH FUTURE RATES?*

3.3.1 "Prediction Is Very Difficult, Especially about the Future"

The title of this section originates from one of my favorite quotes. It is true for life in general and more so for financial markets.† So I am going to borrow it for this essay regarding predicting interest rates. This rendition is even more apt because many, including some prominent economists and investment experts, perhaps for behavioral reasons, actually think it is easy to make predictions about interest rates.

For years, debates about the merits of risk parity have been curiously centered on the inevitability of rising interest rates, even though risk allocation to interest rates is less than half of the total risk in a risk parity portfolio. What are some of the rationales for forecasting higher interest rates? Some critics simply observed that interest rates had been declining for many years and they were very low compared to their historical average. Therefore, they presupposed mean reversion was in order. Others, like the authors of an open letter to Ben Bernanke (*Wall Street Journal*, 2010), just believed that the Fed's QE programs would lead to currency debasement and inflation, and an inevitable increase of interest rates.

These predictions have turned out to be wrong. In the United States, despite a gradual economic recovery and a recent improvement in the labor market inflation has stayed subdued and UST yields have declined rather than risen. In the Euro zone, with more stringent fiscal policy and a less aggressive ECB, economic growth has not fully recovered while inflation has declined significantly. As a result, yields of many Euro zone countries have plunged to levels that are much lower than UST yields.

Human nature craves predictions. For the last few years, predicting rising interest rates based on either naïve chart reading or simplistic extrapolation of monetary action alone seemed deceptively easy and irresistible to some. In hindsight, one can say that these rationales are too simplistic so we won't discuss them further here.

---

* Originally written by the author in October 2014.

† This quote is so famous that it has been attributed to many including Niels Bohr, Yogi Berra, and Mark Twain.

Beyond these subjective calls, did more quantitative techniques for forecasting interest rates give better predictions? Bootstrapping spot yields for different maturities to derive forward rates is a commonly used quantitative approach to model the term structure of interest rates. For the last few years, the yield curves of many government bond markets have been upward sloping, resulting in higher forward rates compared to current rates. This stylized fact has often been offered as a justification of an expected increase in interest rates. It sounds rather analytical and sophisticated. But is this idea correct?

In this investment insight, we shall dispel the notion that forward rates can be considered as an expectation of future interest rates. We show forward rates have a strong upward bias versus current interest rates. As a result, using forward rates as predictions of future interest rates, in general, calls for rising rates. There are conceptual problems with this approach. An empirical analysis of 60 years of yield data in the United States demonstrates these predictions are also naïve and persistently wrong.

3.3.2 Forward Rates

According to fixed income 101, the forward rate is the future yield on a bond. For example, the yield on a 1-year bond 1 year from now is a forward rate. The forward rate is a rate implied by today's yield curve. Let us suppose today's 1-year zero-coupon yield is 0.10% (y_1) and the 2-year zero-coupon yield is 0.50% (y_2).* One can derive the implied 1-year yield 1-year forward ($f_{1,1}$) by linking them together as follows:

$$(1+y_1)(1+f_{1,1}) = (1+y_2)^2. \tag{3.1}$$

Solving for $f_{1,1}$ gives

$$f_{1,1} = \frac{(1+y_2)^2}{(1+y_1)} - 1. \tag{3.2}$$

If we plug in the values for 1-year and 2-year yields, we have $f_{1,1} = 0.90\%$ or 90 basis points.

Note the implied 1-year forward rate 1 year from now is not only higher than the current 1-year yield but also much higher than the current 2-year yield.† Hence, if you believe the forward rate is a forecast of 1-year yields in October 2015, then you might predict a substantial rise of interest rates 1 year from now. We shall discuss the validity of this forecasting method in the remainder of the chapter.

By the way, the fact that $f_{1,1}$ is higher than y_2 is not unique to the example. When the yield curve is upward sloping, that is, $y_2 > y_1$, it would always be true that $f_{1,1} > y_2$. Conversely, when the yield curve is inverted, that is, $y_2 < y_1$, then $f_{1,1} < y_2$. In other words, an inverted curve implies a decrease in the 1-year yield, which is large enough to take it below the current 2-year yield.‡

* These are roughly the yields of the US Treasury zero curve on October 31, 2014.

† It is actually close to current 3-year yield, which was at 0.96%.

‡ This is easy to prove mathematically from Equation 3.1, which states y_2 is a geometric average of y_1 and $f_{1,1}$.

3.3.3 Financial Interpretations of Forward Rates

Equation 3.1 and the derivation of the forward rate $f_{1,1}$ have many financial interpretations. It is worthwhile to review a few of them here since they can provide some perspective on whether there might be some relationship between forward rates and future interest rates.

The first interpretation arises from choices investors would have when "investing for a period of 2 years." One choice is to buy a 2-year discount bond and hold it to maturity. Another choice is to buy a 1-year discount bond and then at the end of the year roll into another 1-year bond. In this scenario, the required yield on the second 1-year bond at the end of the first year, in order to make the two choices equivalent, is the forward rate $f_{1,1}$. When the curve is upward sloping, the forward rate $f_{1,1}$ must be high enough to compensate the yield disadvantage of the first year.

The second, similar interpretation arises from the choices investors have when "investing for a period of 1 year." One choice is to buy the 1-year discount bond and another choice is to buy the 2-year discount bond and then sell it after 1 year. Since the bond is sold before its maturity, it would be sold at a price consistent with the prevailing market yield or price at the time. In this scenario, the required 1-year yield at the end of the first year, to make the two choices equivalent, is the forward rate $f_{1,1}$. When the curve is upward sloping, the forward rate $f_{1,1}$ must be high enough to offset the advantage of the 2-year yield.

In this interpretation, the 1-year bond will mature at its par value of 100 after 1 year while buying the 2-year bond is subject to price volatility at the end of the year. Arguably, the latter should have a higher return for risk compensation. This extra return is often referred to as term premium and it would imply a 1-year yield lower than $f_{1,1}$.

The first two interpretations outlined above are based on a return equivalency paradigm between different investment choices with the same investment horizon. They are used to derive the forward rate but they do not mean the forward rate is a market rate. A third interpretation is market based. Suppose we enter a forward agreement of buying or selling a 1-year discount bond 1 year from now. What should the rate be for the 1-year bond we use today for the forward agreement? It turns out that the rate must be the forward rate $f_{1,1}$. If not, one can build a risk-free profitable arbitrage using a combination of the 1-year discount bond, the 2-year discount bond, and the forward rate agreement via Eurodollar futures or forward starting swaps.

In this case, the forward rate should be reflected in the market place as it represents the yield of a forward rate agreement. But does it have anything to do with the actual 1-year yield 1 year from now?

3.3.4 Forward versus Future

There are reasons to suspect that forward rates are not necessarily unbiased predictors of future interest rates. Here we present some conceptual reasons. We shall present empirical examinations of their relationship in Section 3.3.5.

First, let us go back to the case of the upward sloping yield curve in which $y_2 > y_1$. Then, the 1-year yield 1 year forward is going to be higher than y_2. It is one thing to expect that an upward-sloping yield curve might qualitatively imply an increase in yields. It is quite another to believe that the 1-year yield 1 year from now will always be higher than "the current 2-year yield." In our numerical example, the 1-year yield 1 year forward would be 90 bps, which is 80 bps more than the current 1-year yield.

Second and more generally, an upward-sloping yield curve always implies forward rates that are higher than current rates. Empirically, yield curves tend to be upward sloping most of the time as we shall see later in the chapter. That cannot mean yields are expected to rise most of the time. Most likely, higher yields for longer maturities reflect term premium in exchange for duration risk (the second interpretation) rather than an indication of higher future yields.

This represents a special challenge in the environment of ZIRP, adopted by the monetary authorities of many developed countries. When short-term rates are bounded below by zero, yield curves can only be upward sloping, resulting in forward rates that are persistently higher than the current rates. It is important to realize that this increase in rates is caused by a technicality and it does not reflect the market expectation of future interest rates.

Third, the failure of forward rates as a predictor of future rates is not unprecedented. One example is forward foreign exchange rates. Uncovered interest rate parity implies that future changes in exchange rates are determined by current short-term interest rate differentials between countries. In theory, a currency with a lower interest rate would appreciate against a currency with a higher interest rate by an amount equal to their interest rate differential. However, the forward exchange rate has been proven to a biased or even perverse estimator of the future level of spot exchange rates. In fact, currencies with low interest rates tend to depreciate, rather than appreciate, against currencies with higher interest rates. This "puzzle" is pervasively exploited through currency carry trades.

For example, the JPY has been a low-interest-rate currency for a long time while the AUD has been a high-interest-rate currency. The forward exchange rate between the two currencies has predicted an appreciation of the JPY against the AUD, but the empirical evidence has been the opposite. In other words, the AUD has tended to appreciate against the JPY, in disagreement with the forward exchange rate. The carry trade exploits this phenomenon by taking a long forward position in the AUD/JPY cross rate.

There is a strong similarity between forward exchange rates and forward interest rates. Both are implied rates from current market rates. They are correct rates to use only if we enter a forward rate agreement today. The profitability of FX carry trades demonstrates that no one should use forward exchange rates as a forecast of future exchange rates. Perhaps we should not have high hopes for forward interest rates either.

3.3.5 Empirical Examination of Forward Interest Rate

In this section, we carry out a simple comparison between forward interest rates and future interest rates for 1-year UST zero-coupon bonds. The goal is not to test their statistical relationship rigorously. Rather, it is to assess whether forward rates are reliable predictors of future rates.

We use the Fama–Bliss zero coupon UST data set that goes back to 1952. We shall use nonoverlapping annual observations of 1-year and 2-year yields, that is, y_1 and y_2 to derive

$f_{1,1}$, according to the steps outlined above. We then compare this forward rate to the 1-year yield 1 year later.*

Figure 3.14 plots the 1-year yields at the end of each year, the 2y/1y slope ($y_2 - y_1$), and the increase of the forward rate relative to the current yield. We note the 1-year yield was quite low in the early 1950s and it rose substantially throughout the following decades, reaching a high of about 14% in the early 1980s. It then started its gradual decline for the next three decades, falling to levels below 50 bps from 2008 to 2013. This pattern is true for yields of all maturities.

The solid line in Figure 3.14 shows the 2y/1y slope ($y_2 - y_1$). The difference has been positive for most of the years. This is especially true since the early 1980s when interest rates started their descent. It became inverted only prior to the burst of the internet bubble in 2000 and the credit bubble in 2006. During the first half of the sample, there were more periods of inversion, during the 1960s and late 1970s.

Finally, the bars show the difference between the 1-year yield 1 year forward and the current 1-year yield. Consistent with our analytical reasoning, the bar is always higher than the line when the line is above zero and it is always lower than the line when the line is below zero. In fact, the bar is approximately twice as high as the line.

It is noteworthy that the forward rate has been consistently higher than the current yield since the early 1980s. This is the period of consistently declining yields. This conflict has made the forward rate a poor predictor of the future rate for this period. Prior to the early 1980s,

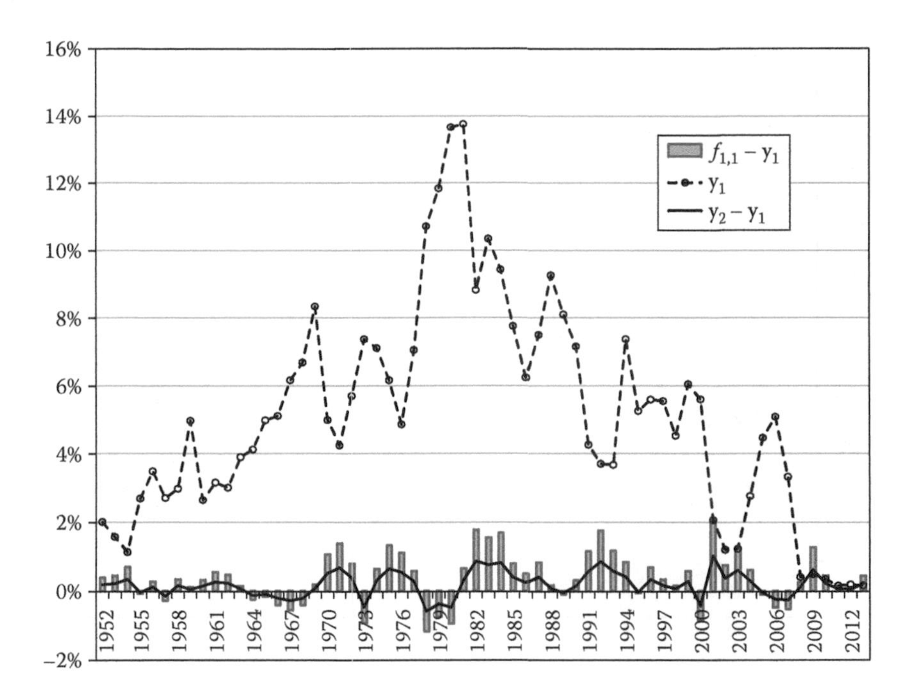

FIGURE 3.14 Annual observation of 1-year yields, and 1y/2y slope, and difference of forward rates and current rates.

* Of course, this comparison can be made for yields of other maturities. We focus on 1-year yields for brevity and the results for other maturities are consistent with 1-year maturity.

however, while the forward rate was still higher than the current yield on average, it was less consistent. However, most problematically, in two periods when the yield rose sharply, once in the 1960s and the other in the late 1970s, the forward rate was below the current yield.

3.3.6 Comparison of Forward Rates and Actual Rates

We now compare the 1-year yield 1 year forward with the actual 1-year yield 1 year later. Figure 3.15 plots the 1-year yield together with the difference between the forward rate from 1 year ago $f_{1,1}(-1)$ and the 1-year yield. If the forward rate is considered as a predictor of the actual rate, this difference measures the forecasting error of the forward rate. When the bar is above zero, the forward rate overstates the actual rate. When the bar is below zero the forward rate understates the actual rate.

Without any statistical test, one can conclude from the graph that before 1981, the forward rate somewhat underestimated the actual rate and after 1981, the forward rate consistently and significantly overestimated the actual rate.

The reason is quite simple: the forward rate consistently predicted higher rates because the yield curve was upward-sloping most of the time. Therefore, when the rate declined in the period after 1981, the forward rate predicted the wrong directional change in interest rates. In the rising interest rate environment prior to 1981, the forward rate prediction was directionally correct resulting in a forecast that was less wrong. However, one must realize that it is not because it had real predictive power.* Rather, similar to the case in which

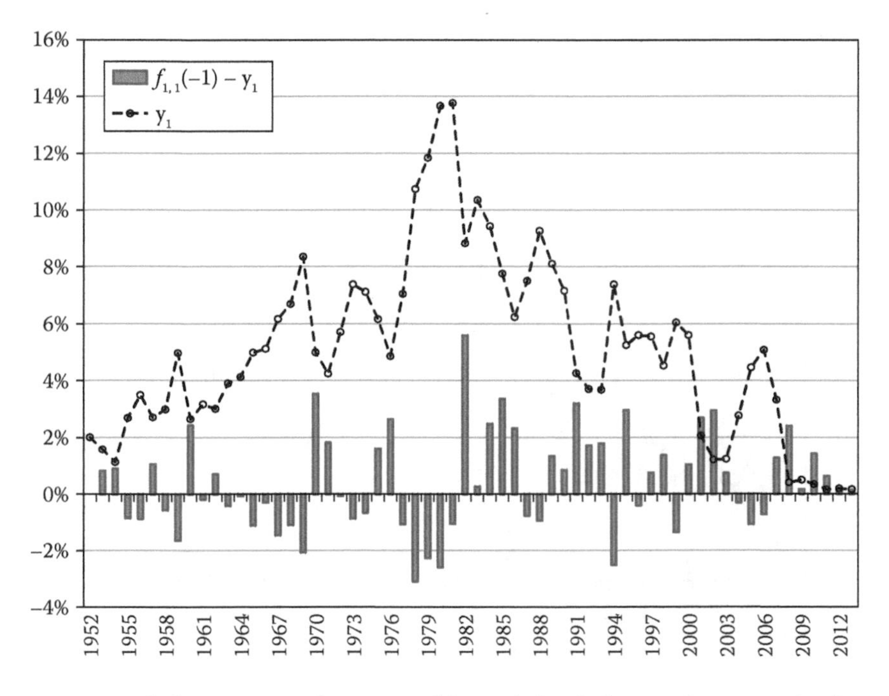

FIGURE 3.15 Annual observation of 1-year yields, and the difference between the forward rates and the actual rates.

* One must go through rigorous tests to tell whether the forward rate has any information for the future rates. Here, we merely point out that the forward rate is not an unbiased estimator of actual rate.

TABLE 3.3 Frequency of Directional Changes Based on Forward Rates and Actual Rates

	Rising (%)	Declining (%)
Forward rate	72	28
Actual rate	51	49

a dead clock being correct twice a day, the forward rate was just being consistently biased upward and got lucky when the actual rate rose.

To examine more closely the structural bias of forward rates, we evaluate the forward rate forecast relative to delivered yields by aggregating annual observations from 1952–2103 into categories of either rising or declining interest rates.

Table 3.3 evaluates the percentage of time the yield curve predicted declining or rising rates compared to what happened in reality. Over this period the forward curve predicted interest rates would decline the subsequent year 28% of the time while increasing 72% of the time. In reality, interest rates declined 49% of the time while they increased 51% of the time. Obviously, expecting rising interest rates nearly three-quarters of the time and only seeing them rise half of the time provides enough empirical support to establish the fact that forward rates have been positively biased estimators of future yields.

Another simple way to evaluate the efficacy of prediction is by hit ratio—the percentage of times when the directional change of the forward rate is consistent with the directional change of the actual rate. Anything less than 50% would indicate a poor performance.

Table 3.4 shows the overall hit ratio as well as the hit ratios for the two subperiods. The overall hit ratio is only 39%. Interestingly, the hit ratio is lower in the first half at 34% than in the second half at 45%. This is because in the first half, the actual rate was rising consistently and the forward rate got it wrong when it was predicting declining interest rates. In the second half, the yield curve managed to predict two recessions but the hit ratio is still below 50% because of its upward bias.

Instead of using forward rates to forecast future interest rates, one could just use current rates as a predictor of future rates. In fact, interest rates, like other financial asset prices, are often modeled, as a first approximation, random walk process in which future "locations" are centered at the current "location."

Figure 3.16 shows the difference between the consecutive yields, which can be viewed as the error of such forecasts. We note there is no persistent bias one way or the other over the entire period. Prior to 1981, there is a slight negative bias since the yield steadily increased, and after 1981, there is a slight positive bias because the yield decreased.

TABLE 3.4 Hit Ratio of Forward Rates to Predict the Change of Interest Rates

	Hit Ratio (%)
1952–2013	39
1952–1981	34
1982–2013	45

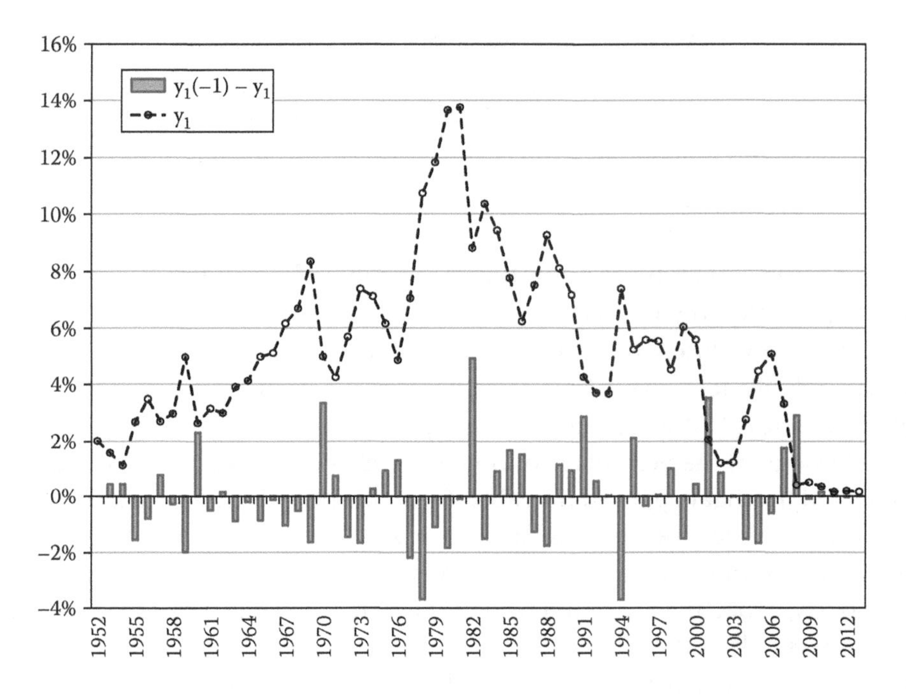

FIGURE 3.16 Annual observation of 1-year yields, and the difference between last year's yield and the current yield.

Comparing Figures 3.15 and 3.16, it can be concluded that forward rates are no better predictors of future rates. In fact, they may be less reliable than current rates—a lesson that many pundits and financial experts could use.

3.3.7 Conclusion

Forecasting is never easy. The experience of the last few years reminds us that it is hard to make predictions about future interest rates based on a simple observation of yield charts or a simplistic extrapolation of market events. In this note, we show that forward rates, often offered as a predictor of future interest rates, are not a reliable predictor either.

The reason is that the yield curve tends to be upward sloping more often than not. This induces the forward rates to have a permanent upward bias relative to current rates, implying a permanent bearishness on the bond market. One can compare this to a permanent bullishness some investors have toward stocks. Both are not useful. Ironically, it is perhaps these permanent biases that have made bonds a better investment over stocks in terms of risk-adjusted return over the last few decades.

In the current market environment where the short-term rates of many developed countries are near zero, forward rates are often significantly higher than the current spot rates. This fact probably has little to do with future changes in interest rates. Rather, it has more to do with the fact that it is impossible to have a flat or inverted yield curve when short-term rates are near zero and investors still expect a term premium.

Of course, nobody should be arguing that interest rates should be permanently depressed at the current level. However, they could stay around the current level for quite some time.

An uninformed investor is better off with using current rates rather than forward rates to form their expectations about future changes in interest rates. Of course, for informed investors, there might be ways to improve upon this baseline assumption, with a better and more rigorous quantitative and fundamental analysis of interest rates, other financial data, and macroeconomic variables.

CHAPTER 4

See the Forest for the Trees

In investing, it is often said that diversification is the only free lunch. The three primary risk premiums that make up a risk parity portfolio not only provide long-term returns but also offer natural diversification benefits to the overall portfolio. These diversification benefits are rooted in macroeconomic fundamentals. Their quantitative manifestation is the low and sometimes significant negative correlations between different risk premiums. Risk parity portfolios utilize positive risk premiums together with these low and negative correlations as ingredients to make that free lunch.

Take stocks and nominal bonds as an example. In a low and stable inflation environment such as the one the world has found itself in for the last decade or so, the dominate factor in determining the returns of both stocks and nominal bonds has been economic growth. Specifically, it is the future economic growth versus the market's expectations. If growth were better than expected, stock markets would usually rise due to better company earnings. Bond yields, on the other hand, would usually rise with bond prices declining, due to the expectation of higher inflation. However, if growth were weaker than expected, the opposite would occur. Stock markets would react negatively to negative growth shocks while bond markets would react positively. Hence, no matter what future economic growth is, stocks and bonds typically provide diversification to each other.

However, this diversification is only real or meaningful when risk allocations to both risk premiums are balanced, as they are in a risk parity portfolio. It is not that there is no diversification in a 60/40 portfolio. There is, but it is minimal because in a 60/40 portfolio, the equity risk allocation overwhelms the bond risk allocation.

Similarly, diversification benefits also exist between inflation risk "premium," and equity and interest rate risk premiums. One only has to go through the same logic process to realize that for instance, nominal bonds and commodities offer diversification against future inflation shocks, whether realized inflation is below or above the market's expectation.

All diversification benefits hinge on the perspective that a properly constructed risk parity portfolio is bigger than the sum of all its parts. One should not miss the forest for the trees by focusing on individual risk premiums in isolation. Unfortunately, many objections from risk parity critics are of this nature. They question the wisdom of including individual asset class/risk premium (e.g., interest rate premium discussed in Chapter 3) in risk parity portfolios based on their views of individual asset classes. Often, this narrow focus on individual asset classes leads to a self-contradictory conclusion. In other words, univariate forecasts for individual assets can become inconsistent with underlying macroeconomic fundamentals.

This chapter provides five investment insights that are related to the issue of diversification in risk parity portfolios. The first one titled "Spear and Shield" uses an ancient Chinese story to illustrate that it is unlikely for both nominal bonds and commodities to simultaneously have negative returns. The second insight extends the theme by pointing out that focusing on individual risk premiums misses the essence of risk parity as a portfolio. The third and fourth investment insights draw on our live experience managing a risk parity portfolio. We show how risk parity portfolios provide protection to investors in the risk-on/risk-off environment after the global financial crisis and how diversification led to positive returns for risk parity portfolios even when bond yields rose significantly in several episodes from 2006 to 2011. The final insight provides a rebuttal to some critics of risk parity by demonstrating that even if we opt to use their own asset return forecasts, portfolios built upon the risk parity principle are superior to traditional asset allocation portfolios.

4.1 SPEAR AND SHIELD*

Learning Chinese is never easy, even for Chinese. However, native students often have an advantage by knowing the stories behind some Chinese words. When I was young, my teacher taught us the following story when the class was learning the word "contradiction." In Chinese, the word is made up by two characters: "spear" and "shield." The story tells why together they mean contradiction.

> During the Chinese Warring States period (476–221 BC), there was a craftsman who made spears and shields and sold them in a market. One day, in the market, he boasted of his spears: "My spear is so strong that it can pierce through anything." Hearing his words, people in the market gathered around his booth. He then continued: "My shield is so solid that no spears can pierce it." A man in the crowd shouted "what if I stab your shield with your spear?"

Some of the criticisms about risk parity somehow remind me of this story. This sounds strange but I will explain this in detail. Several articles have tried to make a case of negative return premiums, that is, excess return over cash, for several asset classes included in risk parity portfolios. One recent article (Inker 2010) argued that two broad asset classes: government bonds and commodities are likely to have negative return premiums.

The arguments against them have a common valuation theme. For bonds, it is rather straightforward or simplistic—the current yield is low according to the historical standard and if interest rates rise in the future, nominal bonds will probably have negative returns. The argument against commodities is a little complex and technical, since commodities

* Originally written by the author in July 2011.

have no value in the conventional sense—they provide neither interest payments nor dividends. However, historically the return of commodities can be decomposed into roll yield and price return. When commodity prices remain roughly the same over time, the roll yield plays a dominant role. Since the roll yield has recently turned negative for many commodities, the expected returns for commodities seem to have dimmed.

While the arguments might appear interesting for individual cases and warrant more in-depth discussion, we address them in separate notes.* However, predicting that both bonds and commodities will have negative returns is self-contradictory. It misses the point of diversification in multiasset class portfolios. The common economic factor underlying both asset classes—future inflation—precludes a simultaneous decline of both. In other words, bonds are the "spear" and commodities are the "shield."

Let us elaborate on this analogy by considering the effect of inflation on these two asset classes. First, the most likely cause of higher interest rates and negative bond returns in the future is clearly higher inflation. However, higher inflation would lead to higher commodity prices resulting in positive commodity returns. Conversely, the most likely cause of lower commodity prices in the future is probably lower inflation or worse, deflation. But disinflation or deflation would likely lead to lower interest rates and positive bond returns. Therefore, from an inflation perspective, bonds win or commodities win, that is, they are natural hedge to each other. Forecasters, like the fine craftsman in the story, should not claim that both assets have negative return premiums.

Historical returns confirm this relationship. Figure 4.1 shows the returns of Treasury bills (T-bills), the Goldman Sachs Commodity Index (GSCI),† and US government bonds

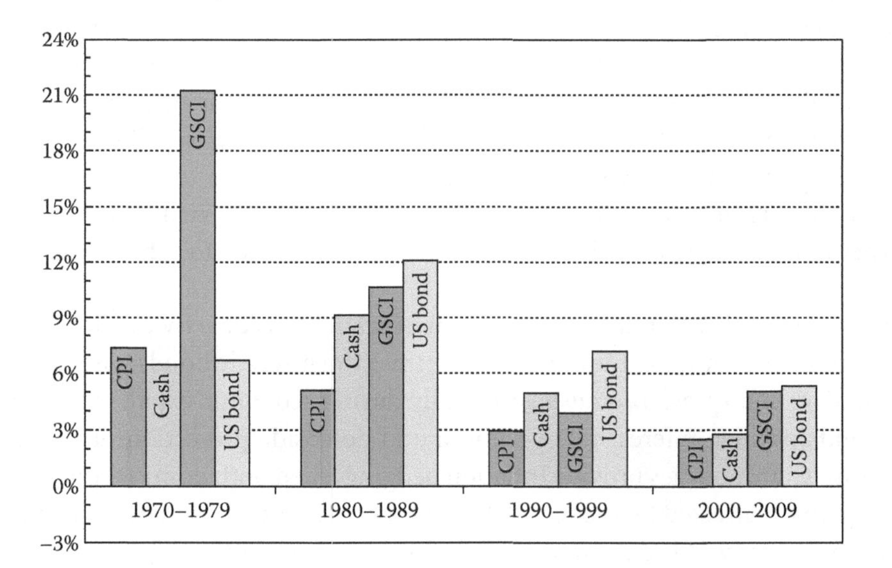

FIGURE 4.1 Annual inflation and asset returns over the four decades from 1970 to 2009.

* We have provided the argument against using roll yield as a predictor of long-term commodity return in Chapter 2 and discussed the fallacy of naïve forecasts of future interests in Chapter 3.

† We use GSCI index for data availability. To hedge inflation, a broadly diversified commodity portfolio is more desirable.

over four decades from 1970 to 2000. In the 1970s, high inflation led to rising interest rates and higher commodity prices. Bonds had almost no return premium over T-bills while the GSCI provided substantial returns over T-bills. In the next two decades, as inflation declined, the opposite happened with the GSCI offering minimal to slightly negative return over T-bills and bonds achieving significant return premium. Therefore, investing in both commodities and bonds has demonstrated that it provides diversification around inflation risk as a source of excess returns.

This relationship only intensified in the most recent period. In the 16 months from January 2010 to June 2011, the return correlation between the GSCI and US government bonds was −0.59. This negative correlation reflects the uncertainty of the future trajectory of both growth and inflation. It is also a manifestation of the "risk-on" and "risk-off" phenomenon of capital markets.

It is likely that the future of interest rates and commodity prices will be driven by the strength of the economic recovery in the developed countries as well as the fight against inflation in the emerging countries. By investing in both nominal bonds and real assets such as commodities and inflation-linked bonds, risk parity aims to diversify inflation risk. In my view, to grow and protect one's investments, investors would arm themselves with both a "spear" and a "shield" rather than none, as suggested by some critics of risk parity.*

4.2 SEE THE FOREST FOR THE TREES†

Risk parity—a portfolio construction approach, when applied to asset allocation portfolios, balances the contribution of the targeted risk premiums from different asset classes. But balancing the risk contribution is not the goal of a risk parity portfolio, but rather a mean to achieving the ultimate objective: a portfolio that achieves stable returns in various economic regimes and across different market cycles. To understand how the risk parity approach can be effective, it is imperative to understand the varying roles that different asset classes are expected to play in order for the entire portfolio to collectively achieve its objective of consistent, risk-adjusted returns. Many critics seem to forget the big picture and instead focus much of their criticism and skepticism on individual parts of the portfolio rather than evaluate how the entire approach is well suited to achieve the investment objective.

The inclusion of three broad asset classes is crucial for risk parity portfolios to achieve consistent returns over time: they are equities, investment-grade bonds, and commodities. Equities deliver equity risk premium, especially during economic expansions, while high-quality bonds provide interest rate premium and downside protection during economic contractions. Commodities provide inflation hedging when both nominal bonds and equities are negatively affected by rising inflation. Once this general framework is understood, we can address many of the common criticisms associated with risk parity. Furthermore, we can ask and address truly important questions regarding risk parity portfolios.

* The experience since the writing of this investment insight has confirmed the thesis of diversification between nominal bonds and commodities. The former has delivered strong risk-adjusted return while the latter performed poorly from 2012 to 2014, partly due to low and declining inflation in many parts of the world.

† Originally written by the author in August 2012.

4.2.1 Miss the Forest for the Trees

One common attack of risk parity is to argue against the inclusion of certain asset classes based on a perceived expectation of its future performance. For example, Inker (2011) uses the fact that the yields on UST bonds had fallen to a very low level, and as a result, may no longer offer any return premium in the future. He is by no means alone in expressing this view. We admit that these are valid observations, but they cannot and should not be the only consideration regarding expected asset returns and investment decisions in the asset class. Depending upon the evolution of the current macroeconomic environment there may be a number of outcomes, such as a sustained economic recovery, or a repeat of the Japan experience, to name two. Asset allocation investors must guard against a deflationary environment in which equities and commodities would perform poorly. On the other hand, if indeed Treasury yields rise due to increases in growth and inflation, equities and commodities in risk parity portfolios would provide upside participation likely more than offsetting the losses from bonds. Again, it is balanced exposure to these asset classes, which provides the balanced return performance across these different environments.

This diversification argument would boost the case for investing in commodities. Rather ironically, in the same article, Inker also makes a case against investing in commodities, based on the negative roll yields of many commodities. However, in Chapter 2 we showed that the roll yield is a poor predictor of long-term commodity returns, much poorer in fact than the bond yield as a predictor for future bond returns, as it is not a traditional valuation measure and it is highly susceptible to short-term supply/demand shocks. Given the possibility of rising as well as falling inflation, it is simply inconsistent or self-contradictory to shun both Treasury bonds and commodities together.

Another common criticism of risk parity is to use simulated risk parity portfolios to demonstrate that it is no better than the traditional 60/40 portfolios in terms of risk-adjusted returns or Sharpe ratios. Examples of this criticism include the studies of Marlena Lee (2011) and Denis Chaves et al. (2011). These studies seem to echo the notion "stocks for the long run" since they show equity risk concentration in association with 60/40 portfolios as not necessarily being an inferior idea for the long run. So are these criticisms right?

The answer is "No." While these studies take a portfolio view over that of the individual assets, a detailed reading reveals that their portfolios are not exactly risk parity. These portfolios either miss important asset classes or are improperly constructed. For example, the analysis by Lee (2011) builds a type of "risk parity" portfolio with equities and long-term government bonds. We note that these portfolios are inferior to true risk parity portfolios on two fronts. First, the absence of any inflation protection makes them susceptible to inflation shocks. Second, the choice of long-term government bonds exacerbates the inflation shock further since long-duration bonds usually suffer the most in a rising yield environment due to rising inflation expectation and in addition, they tend to have the lowest risk-adjusted return across the term structure.

Chaves et al. (2011) analysis includes all three return premiums in their study. However, they make a simple mistake of applying risk parity to an unequal number of asset classes that represent different risk premiums. Among the nine asset classes selected, five are equity-like, one is commodities, and three are bonds. Equal risk allocation to these nine asset classes would obviously result in a greater allocation to equity-like risk and a lower allocation to interest-rate risk. Furthermore, some of the bond asset classes chosen in the study are credit, which are highly correlated with equity risk. It is no surprise that the study finds their simulated portfolios performed to be similar to the traditional 60/40 portfolios. This is because the portfolios they constructed are similar in terms of their risk concentration toward equities. The trees are all there, they are just not planted properly.

4.2.2 Mind the Forest

Rather than focusing on individual asset classes, the focus should be on the total portfolio. The forest can be healthy even if a few trees die or decay, but it could be in danger if all the trees are unhealthy at the same time. Risk parity invests in different asset classes in order to capture different risk premiums. The downside risk is when all return premiums turn out to be negative. In other words, during periods when all asset classes underperform the risk-free rate, risk parity will have negative risk-adjusted returns—an outcome that fails the objective of risk parity portfolios. So how likely is this outcome? The answer to this question is important for understanding the strength and potential weakness of the risk parity portfolios and for building portfolio protection against this risk.

The key metric in addressing the question is the probability that all return premiums of different asset classes are negative over a given time horizon. We choose three asset classes: equities, sovereign bonds, and commodities since they represent three distinct return premiums. They are represented by the S&P 500 index, the 10-year UST bond, and the GSCI commodity index, respectively, and the returns are monthly from January 1970 to July 2012. The risk-free return is the three-month UST bill return. This sample of over 40 years covers periods of both high and low inflation, as well as periods of strong and weak economic growth.

Figure 4.2 shows the discrete probabilities of four different outcomes with respect to the signs of three return premiums over different horizons ranging from 1 month to 120 months, or 10 years. The label NNN denotes three negative return premiums over the risk-free rate, NNP denotes two negative and one positive, and so on.

Several features of the graph are worth noting. First, the probability of triple misses, or three negative premiums (NNN), is consistently the lowest among the four outcomes. It starts at roughly 10% on a monthly basis and declines steadily as the horizon lengthens. It is about 4% over 12 months and 2% over 36 months. The probability stays at a very low level beyond this point, but it does not vanish completely until the horizon reaches beyond 10 years.

Second, the probability of triple hits, or three positive premiums (PPP), increases almost monotonically from about 20% to 60% when the horizon is lengthened to 10 years. This 40% increase comes from the decrease in the probabilities of NNP and NPP. Finally, another notable observation is that the probability of at least one negative return premium, which is equal to the sum of the probabilities of NNN, NNP, and NPP,* is always very high. It is roughly 70% for a 3-year horizon and 50% for an 8-year horizon. This not only highlights

* Or equivalently it is one minus the frequency of PPP.

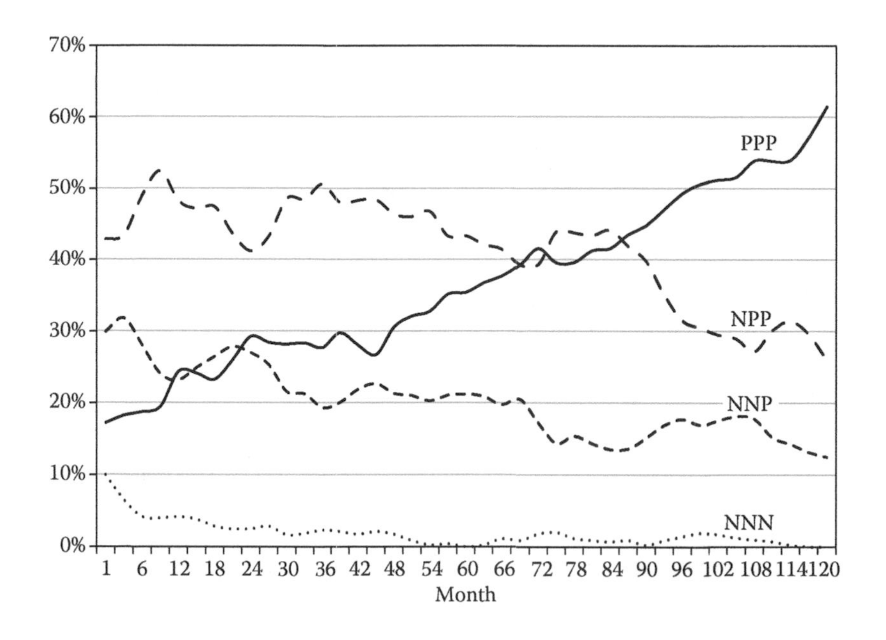

FIGURE 4.2 Probabilities of four distinct outcomes for the excess returns of equities, bonds, and commodities.

the nondiversification risk of a single asset, but also shows how easy it is in hindsight to identify individual assets that might have underperformed for a long period of time.

Figure 4.2 reinforces the case for risk parity as risk premiums of these assets are positive over the long term, and their diversification benefits also increase over the long term with increasing upside probability in PPP and decreasing downside risk in NNN. These statistics, focusing on discrete states of return premiums, are appropriate for risk parity portfolios, which balance the risk and return contributions from different asset classes. The parity feature allows us to treat a hit (P) and a miss (N) equally regardless of asset classes.

On the other hand, for a 60/40 portfolio, because of its risk concentration in equities, a hit (P) or a miss (N) in equities is far more important than a hit (P) or miss (N) in bonds. Therefore, the downside risk of a 60/40 portfolio mirrors closely the downside risk of equities. Figure 4.3 shows the discrete probabilities of experiencing positive and negative equity return premiums. Even though the probability of negative equity excess return declines as the horizon is lengthened, it always stays above 10%, which is much higher than the probability of NNN in Figure 4.2. In other words, the downside risk of three return sources is much less than the downside risk of one.

4.2.3 When and What Do We Worry about?

Even though the probability of three simultaneously negative excess returns diminishes over a long horizon, those outcomes did occur in our sample period. When they happen, it is not necessarily the case that risk parity portfolios would have absolute negative returns. However, it does mean that they, along with all other asset allocation portfolios, would underperform the risk-free asset. Those cases in which investors received negative rewards for taking any risk deserve a closer examination.

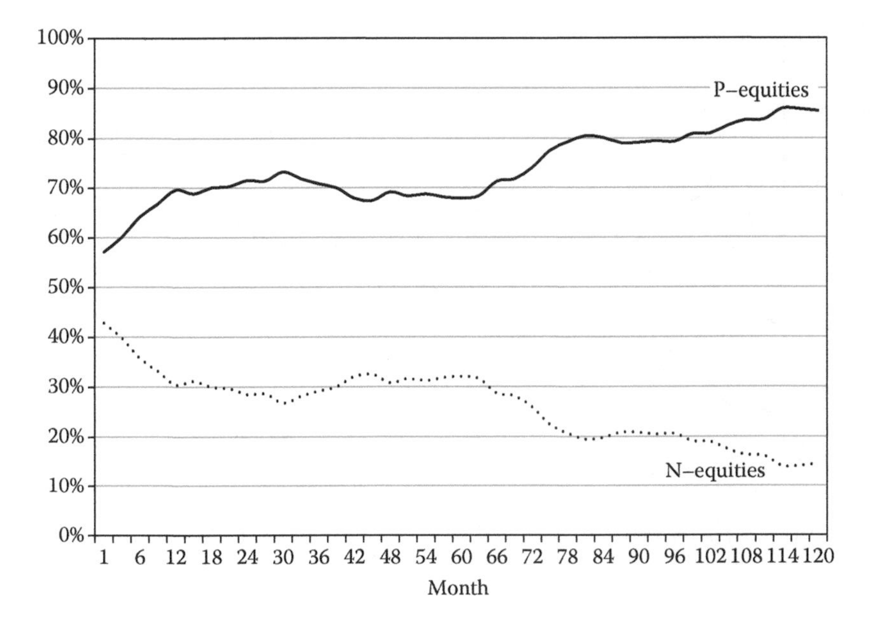

FIGURE 4.3 Frequencies of positive and negative equity excess return over different horizons.

We choose a horizon of 36 months over which to analyze the time series of excess returns from the three-asset classes. Figure 4.4 plots the rolling 3-year annualized excess returns of the S&P 500 index, the 10-year UST bonds, and the GSCI index. To equalize the risk of the three asset classes, we scale the returns of bonds and commodities such that their return volatilities match that of equities. Before we discuss their simultaneously negative excess returns, we make a few remarks about the individual return series.

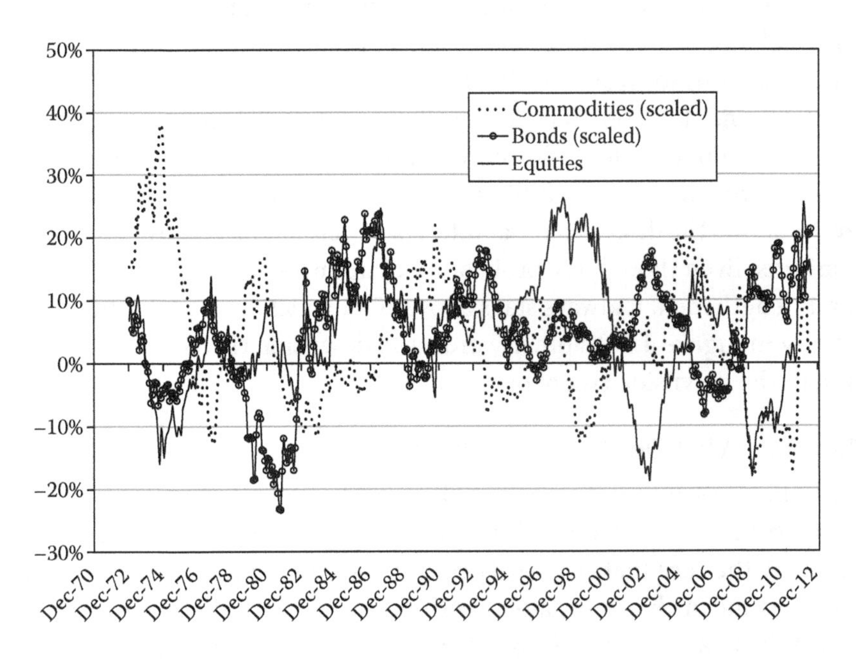

FIGURE 4.4 The rolling 36-month annualized excess returns of three asset classes.

First, after rescaling, the bond return premium shows similar variability to equities. Since the early 1980s, its 3-year excess returns have been quite strong and it was rarely significantly negative, thanks to declining inflation and yields. But it is worth noting that even in the 1970s when inflation rose sharply, there are periods when bonds delivered positive excess returns. Second, the return series of equity premium is volatile, but it predictably follows the business cycles, rising during expansion regimes and falling sharply during recessions. While the returns of equities and bonds more or less offset each other since the year 2000, they are more in-sync prior to 2000. This changing relationship is a reflection of changes in real growth and inflation.

Finally, commodity returns have experienced more sharp drawdowns. Commodities also had the biggest spike during the first oil shock from 1973 to 1974. While they have behaved in a similar fashion to equities since 2002, commodities were a good diversifier to both nominal bonds and equities prior to 2002.

A visual inspection of Figure 4.4 shows that there are more negative asset class returns during the 1970s and early 1980s. To illustrate this more clearly, we first convert the returns for the three asset classes in Figure 4.4 to their signs, 1 for positive and –1 for negative. Next we add the three signs together to get an aggregated time series, which takes on four numerical values: 3 for three positive returns (PPP), 1 for two positives and one negative (NPP), –1 for two negatives and one positive (NNP), and –3 for three negatives (NNN). The dark solid line (with left axis) in Figure 4.5 shows this series.*

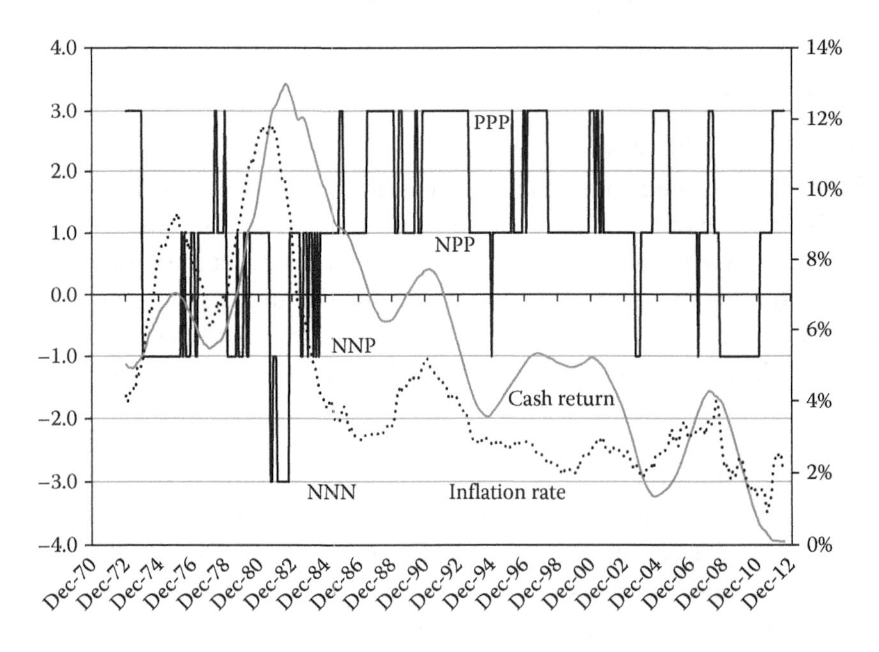

FIGURE 4.5 The solid stepwise line depicts the discrete states of positive and negative return premiums of the three asset classes with 3-year rolling window. The two curves are 3-year rolling annualized cash return and annualized inflation rate with right axis.

* If we compute the frequencies of the four states over the entire period, we arrive at the probabilities of NNN, NNP, NPP, and PPP for the 36-month horizon, which are plotted in Figure 4.2 as four points along the four different curves.

Figure 4.5 makes it apparent that the cases of three negatives as well as most of the cases with two negatives occurred during the 1970s and the early 1980s. The fundamental difference between those early periods and the period since then is the level of inflation. In Figure 4.5, we also plot the rolling 3-year annualized rate of inflation and annualized cash returns (with right axis). The inflation rate rose from near 4% to 9% during the first oil shock, and then to 12% during the second oil shock. These two spikes in inflation devastated both nominal bonds and equities while commodities provided inflation hedging. However, this hedging was not always guaranteed. When commodities failed to provide positive excess returns, the result was the triple miss, where all three asset classes were generating negative excess returns. In summary, the culprit has been inflation shocks, which poses a real threat to all asset allocation portfolios including risk parity.

4.2.4 Defense against Future Inflation Shocks

How serious is this risk today? First, inflation is quite low in the aftermath of the global financial crisis and growth remains weak despite low interest rates. The period of high inflation, should it come, is likely some years away. Second, to a large extent, risk parity portfolios have many inflation hedges built-in by including real assets, such as commodities and inflation-linked bonds, which a typical 60/40 portfolio might lack.

On the other hand, it is never too early to prepare. While risk parity's inflation protection is adequate during periods of moderate inflation, it might not be sufficient for periods of sharp increases in inflation. So how do we shore up inflation defenses under those scenarios? While a full analysis of various approaches is beyond the scope of the present chapter, we think the following list can be useful.

4.2.5 Dynamic Risk Allocation

When inflation is persistently high, the risk-adjusted returns of nominal bonds and equities would likely be lower than that of inflation-linked bonds and commodities. A tactical shift in risk allocation can dynamically allocate more risk to real assets at the expense of traditional assets, such as equities and nominal bonds. Therefore, dynamic risk allocation can improve the overall risk-adjusted return of the total portfolio. Our research has shown that the benefit of dynamic risk allocation would have been quite large in the 1970s (see Chapter 6).

4.2.6 Deleveraging

When all return premiums are negative, any long-only exposure to these asset classes will result in lower portfolio returns. On the other hand, cash returns are often quite high during these periods, due to rising short-term interests, illustrated by Figure 4.5. As a result, another form of inflation protection would be reducing the portfolio's risk exposures such that the total portfolio return becomes more dependent on the cash return, and less dependent on risk premiums. The deleveraging can take place due to rising volatilities and correlations of asset returns. It can also be an active decision based on a systematic process.

4.2.7 Additional Inflation Exposures

If desired, we could also use additional asset classes for inflation protection and increase the allocation to inflation-sensitive sectors within individual asset classes. For example, an equity portfolio with a higher allocation to commodity-related stocks would tend to perform better in a rising inflation environment than a traditional equity portfolio. In a rising inflation environment, a nominal bond portfolio with a greater allocation to the short end of the term structure might outperform a nominal bond portfolio that is heavily weighted to the long end of the term structure.

4.2.8 Conclusion

This research note addresses most criticisms aimed at risk parity portfolios. Such criticisms are either myopically focused on individual asset classes thus missing the point of portfolio diversification, or are based on improper interpretation and implementation of risk parity.

When constructed properly, the potential weakness of risk parity does not lie in the performance of individual assets. Rather it is when all assets have negative return premiums. We conduct a historical examination of three asset classes since 1970, and show that this scenario has very low probability of occurrence and would be most likely during periods with persistently high inflation. There are a number of ways to provide additional inflation protection for risk parity portfolios. The first is dynamic risk allocation, which can tilt portfolios more toward real assets. The second is deleveraging to cut portfolio exposures or portfolio risk. Finally, another approach is to build customized exposures to individual asset classes, such that they would be more sensitive to inflation.

4.3 RISK-ON RISK-OFF AND RISK PARITY*

Every financial crisis gives rise to its own catch phrases and acronyms, and the crisis of 2008 is no exception. "Subprime" and "shadow banking" share the blame for causing the crisis. QE and ZIRP describe the monetary actions designed to fight the crisis. Now, almost 4 years later, "risk on risk off" is becoming the most commonly used investment jargon to describe the condition of the global financial markets in the aftermath of the crisis. In fact, as a testament to its indoctrination into the lexicon of investing it has even been assigned its own acronym—"RORO."

What exactly is risk on risk off? What are the symptoms and the underlying causes of this phenomenon? What are its effects on investment portfolios and how do investors mitigate the negative effects of risk on risk off? In this research note, I address these questions from the perspective of an asset allocation portfolio. We have found that because of changes in correlations between different assets, investors need to pay closer attention to risk management in asset allocation portfolios. In addition, we suggest that while risk parity strategies are well suited for the macroeconomic environment that underlies the risk on risk off phenomenon, it is important to consider its impact on some asset classes such as commodities and corporate bonds in a risk parity portfolio.

4.3.1 Risk On Risk Off

Risk on risk off refers to financial market behavior where almost all risky assets such as equities, commodities, low-grade corporate, low-grade sovereign credit, and risky currencies

* Originally written by the author in May 2012.

tend to move together in tandem. Moreover, the change of direction between on and off happens in much shorter time periods, induced by frequent macroeconomic events across the globe.

In contrast to risky assets, safe assets including high-quality sovereign bonds such as US and German government bonds, low yielding currencies such as the USD, JPY, and for a while the Swiss franc and gold, behave as *safety off safety on* as opposed to risk on and risk off. These seemingly capricious and dramatic shifts in investors' risk tolerance have become an even more dominant driver of asset pricing than the fundamentals of the individual markets themselves.

At first glance, risk on risk off appears surprising and even extreme. But it is not a new phenomenon at all. To some degree, financial markets have always behaved this way. Most commonly, markets are driven by business cycles, which periodically undergo expansionary and recessionary regimes. As the world economy and the global financial markets became more and more integrated the risk on risk off phenomenon has taken on a global form instead of being country specific. For example, the markets transitioned from a risk on period during the equity rally of the late 1990s to a risk off regime when the tech bubble burst and a mild recession occurred from 2000 to 2002. Another long risk on cycle occurred from 2003 to 2006 only to be followed by the Great Recession and the global financial crisis, which triggered a long and painful risk-off episode. Perhaps if the Great Recession had been a normal business recession and the world economy had recovered strongly afterwards, we would have transitioned back into a period of risk on and the acronym RORO may have never been born.

Unfortunately, this was not the case. Instead, the global economic recovery remains timid despite QEs and ZIRPs. To complicate matters, we have nearly exhausted available monetary stimuli, fiscal policies are constrained by heavy government indebtedness, and Euro-zone countries are adversely affected by both the common currency and fiscal austerity. There exists little agreement among policymakers, central bankers, and economic experts as to how to get us out of the current economic predicament. The world economy is at a cross road—the best scenario appears a continued muddle-through and the worst case would be a global version of Japan's experience of lost decades. Such uncertainty seems to be fertile ground for perpetuating risk on risk off.

Nevertheless, there are some new characteristics of this round of risk on risk off that are different from previous experiences. First, there has been broader participation of asset classes and subasset classes. Second, as mentioned above, the frequency of on and off cycles has been much higher. Since 2009, markets have had significant risk on risk off episodes every year and often they can been seen on a daily basis. Later in the chapter, we shall discuss in detail the underlying causes of risk on risk off, but first we will present some statistical evidence on asset return correlations.

4.3.2 Asset Correlations under Risk On Risk Off

Another catch phrase describing risk on risk off phenomenon is "all correlations go to one." This is obviously an exaggeration but it does speak to the heightened correlations among risky assets.

TABLE 4.1 Correlation Table for Three Asset Classes Represented by Barclays US Treasury Index, S&P 500 Index, and GSCI

	UST	S&P 500	GSCI
UST	1	−0.58	−0.53
S&P 500	−0.58	1	0.71
GSCI	−0.53	0.71	1.00

4.3.3 Broad Asset Class Correlations

We first consider the correlations among three broad asset classes: high-quality government bonds, equities, and commodities. Table 4.1 lists the correlations based on monthly returns of the last 3 years from May 2009 to April 2012. The correlation between the two risky assets, stocks and commodities, is 0.71 while the correlations between the risky assets and the safe assets are significantly negative: −0.58 for stocks and −0.53 for commodities.

Are these correlations normal from a historical perspective? To answer this question, we extend the three pairwise correlations to the early 1980s. These results are shown in Figure 4.6.

It is clear that the recent correlation structure is quite unique in several aspects. First, the correlation between commodities and stocks had never been significantly positive prior to the 2008 financial crisis, but now it is at an historically high level and it is only slightly lower than correlations among equity asset classes (shown later in the chapter). Second, the correlation between stocks and USTs is also at its lowest level of the period and it even exceeds the depths reached during both the heights of the 2008 financial crisis and the bursting of the dotcom bubble in 2000. It is worth noting that the stock/bond

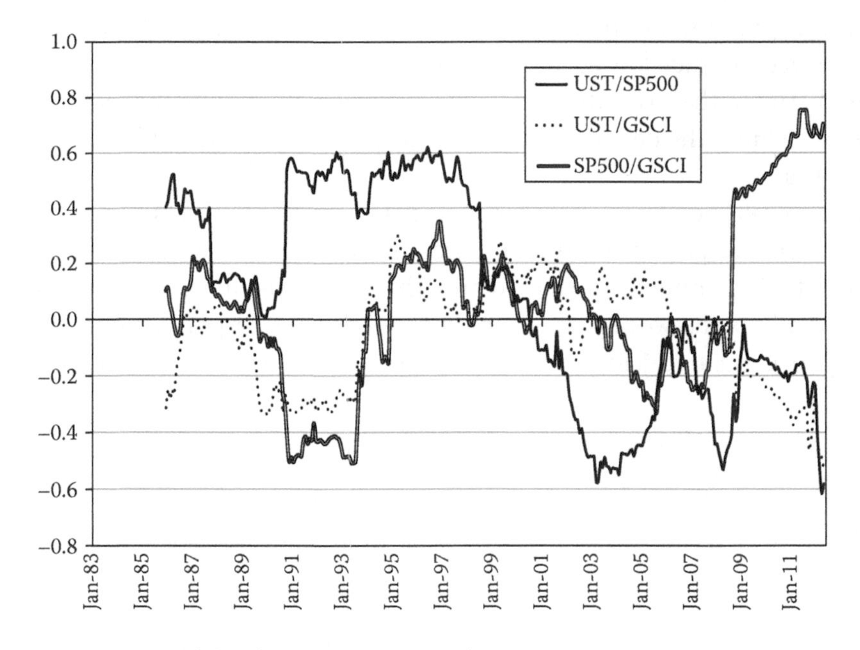

FIGURE 4.6 Time series of the three pairwise correlations.

correlation has been consistently negative for the last 10 years. Finally, the negative correlation between USTs and commodities is also at its lowest level, surpassing the level during the 1980s and early 1990s when inflation was much higher.

However, the dynamics of these three correlations in the most recent period is distinctly different from the dynamics of the disinflation period of the 1980s. In the earlier period when inflation fell persistently, Treasury yields fell and stocks rallied, resulting in a positive correlation between them while their correlations to commodities turned negative. On the other hand, the recent risk on risk off behavior is not driven by surprises in inflation since Treasury yields continued to fall while inflation expectations remain stable. Rather, it has been driven by changes in growth expectations, which have fluctuated between hopes of a strong economic recovery and fears of a deflationary, double-dip recession.

4.3.4 Correlations of Fixed-Income Assets

The correlations within investment grade fixed-income markets show a similar risk on risk off dynamic, in which riskier assets became more correlated with stocks and less correlated with safe assets.

Table 4.2 shows the most recent 3-year correlations among USTs, US investment grade corporate bonds, and international government bonds. The correlation between the United States and the international government bonds is 0.62 while the correlations between the corporate bonds and the other two are much lower at 0.28 and 0.35, respectively. These correlations, especially the latter ones, have been affected by the risk on risk off behavior.

Figure 4.7 shows the time series of these three correlations. Prior to the financial crisis, UST bonds and US corporate bonds used to have the highest correlation at a value above 0.9. But this correlation dropped precipitously during the financial crisis and still remains at a low level. The correlation between US corporate bonds and international sovereign bonds show a very similar change. Throughout the previous almost 30 years, corporate bonds had never had such low correlation with government bonds. On the other hand, the correlation between the two safe assets remains very high, despite the fact that the WGBI ex US index includes the debt of some of the Euro zone countries embroiled in the sovereign debt crisis.

While corporate bonds are now less correlated with safe assets, its correlation with stocks has risen sharply during the financial crisis, and it remains significantly positive today, due to their common exposure to growth risk. This is shown in Figure 4.7 by examining the correlation between changes in the United States corporate bond yield spread to the US 10-year Treasury yield and the S&P 500 index returns. We have used the negative

TABLE 4.2 Correlation Table for Three Fixed-Income Asset Classes Represented by Barclays US Treasury Index, WGBI ex US, and Barclays US Corporate Index

	UST	WGBI ex US	Corp
UST	1.00	0.62	0.28
WGBI ex US	0.62	1.00	0.35
Corp	0.28	0.35	1.00

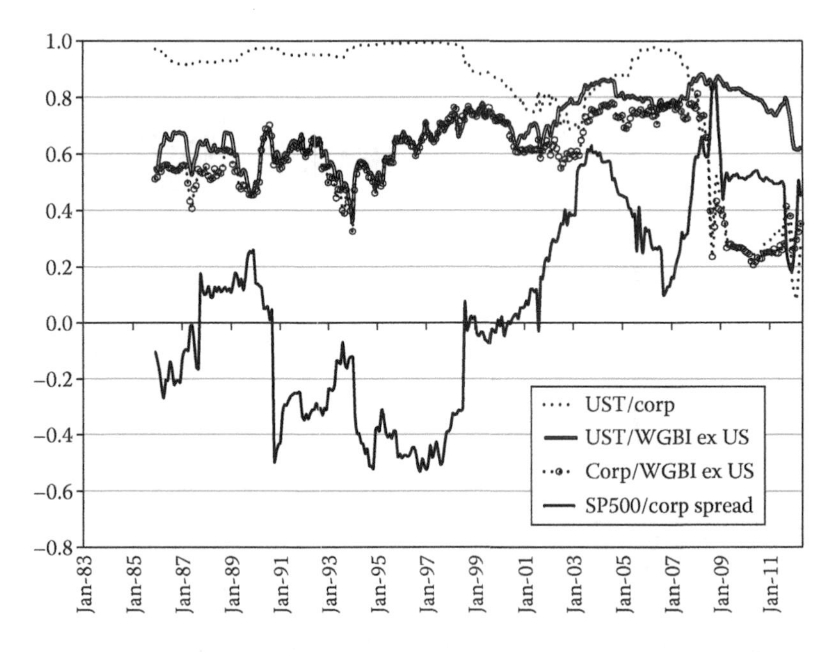

FIGURE 4.7 Time series of the three fixed-income correlations and correlation between corporate bonds and stocks.

of the yield change so it represents the corporate credit premium to Treasury. This correlation used to be slightly negative prior to 2000 but it is now significantly positive. In other words, corporate premium is now highly correlated with the equity market. When financial markets experience risk on risk off, assets with common growth risk exposures will have higher correlation even if they might normally be considered as different asset classes.

4.3.5 Correlations of Equity Asset Classes

The old Wall Street adage "Diversification often disappears when needed most" describes the behavior of correlations among equity asset classes rather accurately during the financial crisis. This has also been true in the recent period of risk on risk off.

Figure 4.8 plots the three pairwise correlations among US equities, international equities, and emerging market equities. First, we notice the overall level of correlations has gone up during the last 10 years. Second, the correlations typically shoot up significantly when equity markets fall, whether it was the 87 crash, or the bursting of the tech bubble, or the 2008 financial crisis. Finally, the correlations have stayed at their elevated levels for the last 3 years. The risk on risk off has been a global phenomenon and therefore diversification across the country dimension within equities has been of limited value to an asset allocation portfolio.

4.3.6 Correlations of Commodities

Commodities used to be a diverse group but for the last 3 years we have also seen increased correlations among different commodities, especially those related to global economic growth such as commodities in the energy and industrial metals sectors.

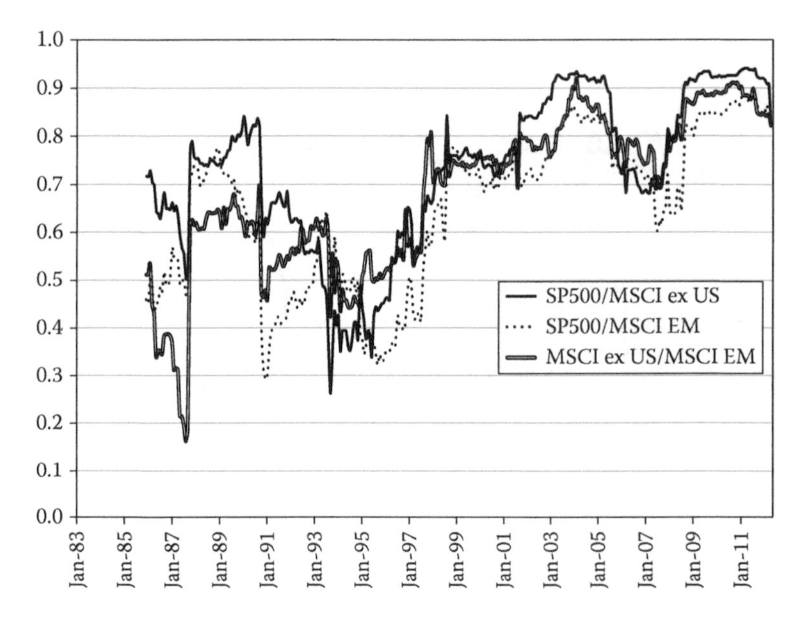

FIGURE 4.8 Time series of the correlations among the S&P 500 index, MSCI ex US index, and MSCI EM index.

Figure 4.9 displays the 3-year rolling correlations among the energy, industrial metals, and precious metals sectors. The correlation between the energy and industrial metals sectors, which are growth sensitive, tends to rise around the time of a recession. Currently, it has risen to its highest level over the past 30 years. In contrast, the correlations of precious metals to the other two sectors have dropped significantly during and after the financial

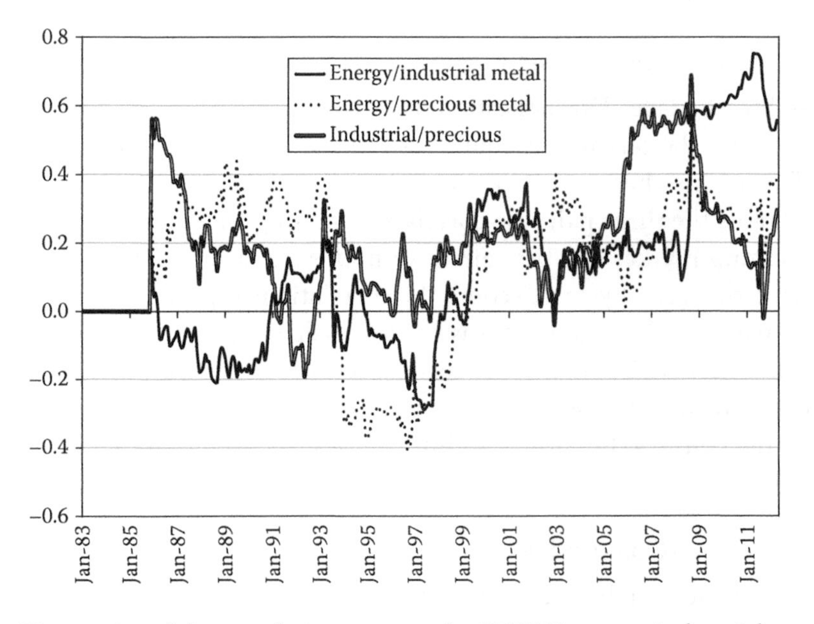

FIGURE 4.9 Time series of the correlations among the DJ/UBS energy, industrial metal, and precious metal indices.

crisis due to the performance of gold as a safe asset. However, in the most recent period, these two correlations started to climb as gold has temporarily lost its standing as a safe haven asset.

As we have discussed earlier, commodity exposure achieved by conventional indices has not been a diversifying asset class in a risk on risk off environment. This is because these indices are typically dominated by growth-sensitive sectors such as energy and industrial metals. Other commodities such as those in the grain and soft sectors have some degree of growth exposure embedded in them as well. In this context, they are analogous to corporate bonds. Therefore, in the environment of risk on risk off, commodity exposures achieved using traditional production or trade-weighted indices offers limited value in terms of diversification.

4.3.7 Causes of Risk On Risk Off

Whether the phenomenon of risk on risk off will continue is an open question. In order to answer this question, we must have a general understanding of the causes of risk on risk off. Some speculate that this has become the new norm of financial markets after decades of global integration. Others attribute it to the ease of trading vast baskets of securities and asset classes with an ever-expanding universe of exchange-traded funds and notes.

While these technical reasons might have played a role, we believe the causes are more related to the macroeconomic environment of weak growth and low inflation that are the results of the financial crisis. Anecdotally, over the last few years, violent episodes of risk on and risk off have been triggered by two intertwined types of macroeconomic events. The first is either renewed hope or concern about global growth and the second is either the perceived resolution or the renewed fear of the sovereign debt crisis. The relationship is that stronger growth would ease the debt burden of governments, corporations, and consumers, which in turn fosters further economic growth. On the other hand, slower growth would worsen the debt problem, which further weakens aggregate demand. The interaction of the two can create a feedback loop that greatly affects the future of the global economy. The risk on risk off phenomenon is one way in which financial markets price in such bimodal outcomes.

This explanation is reinforced by the experience of Japan during the last two decades. Of course, there are significant differences between that localized event and the current global environment. Regardless, the similarity is equally strong in terms of their financial crisis and the prolonged deleveraging process afterwards. As we shall illustrate in Chapter 6, during the ensuing two decades after the Japanese equity and real estate bubble, there has been and still is persistent negative correlation between the returns of Japanese equities and Japanese government bonds—a country version of everlasting risk on risk off.

Another potential cause, especially related to commodities, is the compound risk of growth and inflation. In normal market and economic periods, real growth and inflation are often uncorrelated. But financial crises are usually deflationary and in a low inflation environment, causing growth risk and inflation risk to overlap. In other words, if the economic recovery strengthens then inflation tends to rise. This relationship is further strengthened by central bankers (except for the ECB perhaps) that might tolerate above-average inflation to ensure a durable economic recovery. In the case of a faltering recovery, inflation would likely decline or potentially evolve into deflation. This scenario could explain the positive correlations among equities, commodities, and the break-even yields of inflation-linked bonds.

Another possible explanation of risk on risk off is that it is associated with liquidity flows provided by the Fed's quantitative easing programs. The added liquidity from the Fed's asset purchases has sown the seeds of the rally in risky assets but when the QEs stop, liquidity dries up causing risky assets to decline. This is certainly plausible, but it is not necessarily an independent factor because the Fed's QE programs have followed the ebbs and flows of the economic data in terms of both growth and inflation. Perhaps the best way to describe its effect on risk on risk off is that it has accentuated the magnitude of asset price changes in the on/off cycle.

If these are the fundamental reasons for risk on risk off, then there is a high probability that the phenomenon will continue. The deleveraging process takes a long time, and it naturally causes investor sentiment to swing between optimism and despair, that is, risk on and risk off.

4.3.8 Implications to Asset Allocation Portfolios

Since risk on risk off manifests itself in almost all asset classes and more importantly across different asset classes, its impact is likely most significant for asset allocation portfolios. Here, we discuss its implications to traditional balanced portfolios and risk parity portfolios.

We first discuss the impact on portfolio risks that are directly related to the changes in correlations that we attribute to a risk on risk off environment.

4.3.9 Traditional Balanced Asset Allocation Portfolios

Traditional 60/40 portfolios have a majority of their risk in risky assets. Needless to say, they are generally more volatile during risk on risk off regimes. A major contributor to this volatility jump is due to its concentration in equities, but even the composition of the portfolio's 40% fixed-income exposure can have a meaningful impact on its risk profile. Qualitatively, if the allocation is in safe government bonds, then the total portfolio risk will be lower, due to its negative correlation with equities. In contrast, if the allocation is concentrated in corporate bonds, the increased correlation between credit spread and equities in a risk on risk off environment could cause the portfolio risk to rise significantly.

To see this qualitatively, we construct two 60/40 portfolios with the same equity investments in the MSCI world USD hedged index. For the 40% fixed income, the first portfolio invests 20% in the Barclays US Treasury index and 20% in the WGBI ex US index while the second portfolio invests the entire 40% in the Barclays US corporate index.

Figure 4.10 plots the rolling 3-year annualized volatility of the two portfolios and their difference. While both portfolio risks oscillate along with equity market volatility, the difference is quite small prior to 2008. After the financial crisis, the volatility difference increases above 2% and is still high following a recent drop. The lesson is that one needs to reduce corporate bond exposure to control the total portfolio risk of a 60/40 portfolio.

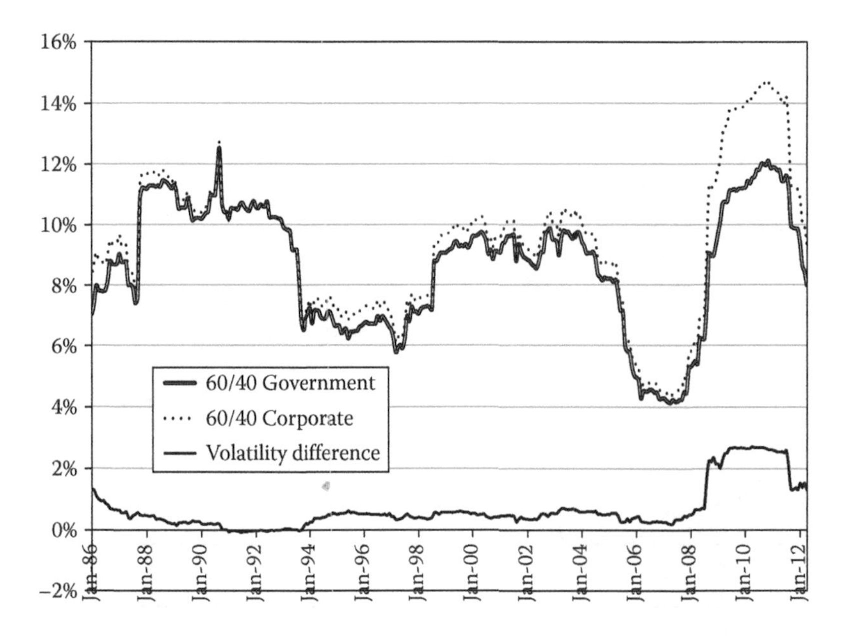

FIGURE 4.10 Three-year rolling annualized volatility of two 60/40 portfolios and their difference: 60/40 Government consists of 60% MSCI world, 20% Barclays US Treasury, and 20% WGBI ex US; 60/40 Corporate consists of 60% MSCI world and 40% Barclays US Corporate.

4.3.10 Risk Parity Asset Allocation Portfolios

Almost by design, risk parity portfolios are less sensitive to risk on risk off since they balance risk contribution from risky assets and low-risk fixed-income assets. However, not all fixed-income assets are alike and how we allocate within the fixed-income asset class presents the same challenge as in the 60/40 portfolios. In a risk on risk off environment, credit risk premium and especially default risk premium become highly correlated to equity risk premium. We should take this into account when managing risk parity portfolios' aggregated exposure to equities and corporate bonds.

To illustrate the impact of the fixed-income allocation on portfolio risk, we construct two simple proxies of risk parity portfolios, both of which invest 40% in the MSCI world index and 160% in bonds. The first invests 80% in the US Treasury index and 80% in WGBI ex US index, while the second cuts those in half and invests 80% in the US Corporate index. Figure 4.11 plots the rolling 3-year annualized volatility of the two portfolios and their difference. The first noticeable feature is that both volatilities actually declined prior to the 2008 financial crisis, partly due to the increasingly negative correlation between stocks and bonds. Second, the volatility difference between the two portfolios was stable and limited prior to the financial crisis. However, similar to the 60/40 portfolios, the volatility difference became substantially higher as the 2008 crisis began. The risk parity portfolio with half of the fixed-income exposure allocated to corporate bonds had much higher portfolio risk because of the high correlation between corporate bonds and equities. Failure to take this into consideration will dilute the diversification benefit of risk parity portfolios.

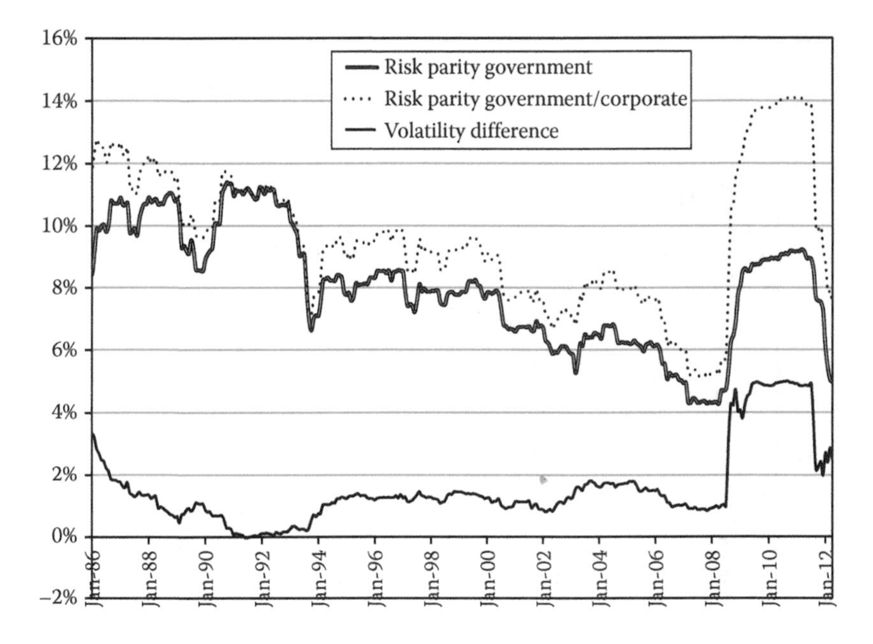

FIGURE 4.11 Three-year rolling annualized volatility of two risk parity portfolios and their difference: Risk Parity Government consists of 40% MSCI world, 80% Barclays US Treasury, and 80% WGBI ex US; Risk Parity Government/Corporate consists of 40% MSCI world, 40% Barclays US Treasury, 40% WGBI ex US, and 80% Barclays US Corporate.

Risk parity portfolios in practice also often include commodities as an asset class for inflation protection. The increased correlation between commodities and equities presents additional issues for the risk management of risk parity portfolios.

To illustrate this point, we construct a third risk parity portfolio by including commodities and changing only the risky assets in the risk parity portfolio with government and corporate bonds. Instead of a 40% allocation to the MSCI world index, this portfolio split the allocation equally between the MSCI world index and the GSCI. Normally, one might expect the portfolio including commodities to be more diversified and to have lower portfolio risk. Figure 4.12 shows that is indeed the case prior to 2003. After 2003, the supposedly more diversified portfolio actually has had higher risk than the portfolio without commodities. It is true that in a portfolio with multiple assets, the total portfolio risk depends on many inputs. The heightened correlations among risky assets—equities, commodities, and corporate bonds—certainly played a role. This is just another example that more assets does not necessarily translate into more diversification in a risk on risk off environment.

4.3.11 Return Implication of Risk On Risk Off

While the impact of risk on risk off on portfolio risk is clear, its impact on future returns of various assets is more complex. We can think of this problem in term of different horizons. First of all, over the short term, risk on risk off necessarily means the returns of risky assets and safe assets are opposite of each other. But this does not imply that over the intermediate term, the two sets of returns will be widely divergent since the average returns in the

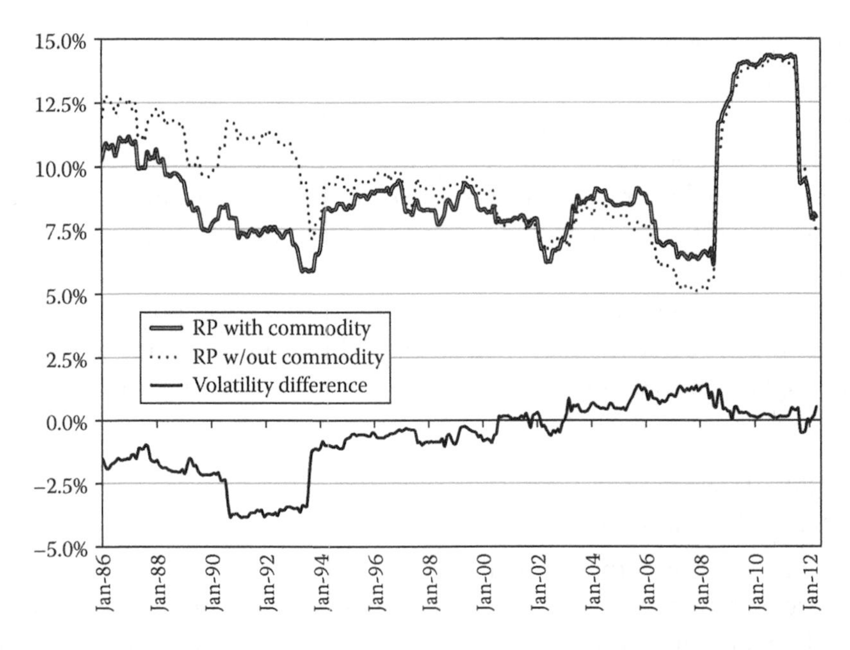

FIGURE 4.12 Three-year rolling annualized volatility of two Risk Parity portfolios and their difference: Risk Parity without commodities consists of 40% MSCI world index, 40% Barclays US Treasury, 40% WGBI ex US and 80% Barclays US Corporate; Risk Parity with commodities substitutes 20% of MSCI world index with 20% GSCI.

future are not solely determined by correlation. For instance, over the last 3 years, asset price reflation has occurred in almost all asset classes.

But one cannot help thinking that the situation could be rather different over the long term. For example, under the bimodal outcome of either stronger economic recovery or economic stagnation or deflation, the returns shall be very different for risky assets than they will be for safe haven assets, that is, their returns will be widely opposite of each other. From the perspective of risk management over the long term, risk parity portfolios will be less sensitive to the bimodal outcome than the traditional 60/40 portfolios.

4.3.12 Conclusion

The current environment presents many challenges to investors. The private sector is deleveraging, the public sector is saddled with unprecedented levels of debt and the world economy is experiencing low inflation and subpar growth. The uncertainty around whether global growth will be sustainable to service increasing levels of debt continues to foster a risk on risk off environment where all risky assets with exposure to growth risk move in tandem. This is consistent with the observation that we make in this note that correlations among various assets change drastically in a risk on risk off environment with correlations increasing among risky assets and correlations become more negative between risky assets and safe assets. This results in drastic changes in the level of portfolio risk in asset allocation portfolios. One example of how changes in the correlation structure can impact a portfolio's volatility is the impact of an allocation to credit risk premium in an equity-centric asset allocation portfolio. As the correlation between credit risk and equities increases during a risk on risk off environment, the portfolio's total risk increases as the equity risk concentration grows.

These correlation changes have significant impact on risk parity portfolios as well. We believe that risk parity portfolios that balance the risk allocation from various asset classes or risk premiums, are better at navigating a risk on risk off environment. However, since several risk premiums such as equities, credit, and commodities are now highly correlated, a portfolio with too much risk allocation to these risks can become heavily tilted to growth risk, and to some extent, similar to the traditional 60/40 portfolios. Therefore, in order to remain truly diversified in a risk on risk off environment, one needs to take into account the higher correlations among risky assets when building and managing a risk parity portfolio.

4.4 THE RISK PARITY CONUNDRUM: RISING RATES AND RISING RETURNS*

The simple bond math says that a bond portfolio loses value when interest rates rise and common sense dictates that a levered bond portfolio loses even more value when interest rates rise. While both the simple math and the common sense are often correct, the popular notion that a risk parity portfolio is also destined to lose value in a rising interest rate environment is unsubstantiated. In fact, our experience has been quite the opposite.

We have been running risk parity portfolios for clients since 2006. During that period, even though interest rates have declined in general due to the Fed policy, weak economic recovery, and low inflation, we have actually experienced five different episodes of rising interest rates as listed in Table 4.3.

Over these subperiods, interest rates increased nearly 100 basis points on average while risk parity delivered an average annualized return of 7.50%.† The success of risk parity in rising interest rate environments should not come as a surprise to those who understand risk parity. Risk parity portfolios are designed to capture market risk premium efficiently in a variety of different market cycles including rising, declining, and range bound interest rate environments. In this note, we seek to demystify the relationship between rising interest rates and a properly constructed risk parity portfolio by making three points. First, it is

TABLE 4.3 Episodes of Rising Yields (in Basis Points) and Returns (Annualized) of Asset Classes and Risk Parity Portfolios

Period	Δ10 y	Δ5 y	S&P 500 (%)	US WGBI (%)	Risk Parity (%)
01/06–06/06	+75	+74	3.54	−2.50	−5.89
12/06–06/07	+57	+48	26.44	0.31	0.96
04/08–05/08	+65	+99	40.83	−16.23	17.56
01/09–12/09	+163	+110	23.45	−3.69	7.16
09/10–03/11	+100	+90	49.32	−4.67	17.60
Average	+92	+88	28.72	−2.78	7.50

* Originally written by the author in September 2012; coauthored with Bryan Belton.

† All returns are annualized to compensate for differences in the length of different periods.

important to understand that the simple bond math is not always right, bonds do not automatically decline in value in a rising interest rate environment. Second, it is important to understand that a well-constructed risk parity portfolio should not allow the contribution to return from any asset class including, but not limited to fixed income, to have an undue influence on the entire portfolio. Other asset classes within risk parity portfolios are expected to provide positive returns in a rising interest rate environment. Finally, we believe that well-designed risk parity portfolios have a systematic flexibility built into its investment process to allow the portfolio to be adaptive to changes in market cycles of interest rates.

4.4.1 Do Not Let Bond Math Fool You

Duration is the most common measure of a portfolio's sensitivity to changes in interest rates. Duration can be used to approximate a bond portfolio's price return due to a change in interest rates: $r = -D \cdot \Delta y$. This expression suggests that a bond with a duration of 10 years would lose 10% of its value for a 100 basis point increase in interest rates. While this expression accurately captures the inverse relationship between bond prices and bond yields, it does so only for parallel changes in yields that occur as the result of an instantaneous shock in the yield curve.

Experience tells us that changes in interest rates of such magnitude rarely occur instantaneously. Rather, they typically occur across a cycle that will last weeks, months, or years. When interest rates rise over time, one must factor in the slope of the yield curve to estimate the likely impact on bond returns. For example, a steep yield curve is indicative of a bond market that has priced in either a large-term premium or a dramatic increase in interest rates in the future. To compensate investors for rising yields, forward yields provide a buffer to protect investors from the impact of their future cash flows being discounted at higher yields. This "buffer" means that not only are bond investors not guaranteed to lose money in a rising interest rate environment, but they can also actually earn positive returns as long as spot yields increased less than what was priced by the forward yield curve.

Table 4.4 shows the term structure of interest rates for a theoretical yield curve, which is close to the UST curve at the time of this writing. While the yield levels are extremely low relative to their historical standards, the slope of the curve is reasonably steep. The steepness in the slope of the yield curve could be indicating that the market is pricing in an aggressive increase in interest rates in the future. Bond markets price in a "buffer" for future increases in interest rates by offering forward yields that are substantively higher than spot yields. As a result, the determining factor of whether a bond's return is positive or negative is not whether interest rates rise or fall, but rather whether interest rates rise more or less than the level of forward yields implied by the term structure of interest rates. Using the theoretical yield curve summarized in Table 4.4, we can use a bootstrapping technique to derive forward yields. An investment in a 10-year bond can be decomposed

TABLE 4.4 Term Structure of Interest Rates and 5-year yield 5-year Forward

	3 M	2 Y	5 Y	10 Y	5 × 5 Fwd
Yield	0.05%	0.25%	0.50%	1.75%	3.02%

TABLE 4.5 Return of a 10-Year Zero-Coupon Bond under Different Interest Rate Scenarios

Δ5y Years from Today	Return (%)
Unchanged	16.01
+50 bps	13.17
+100 bps	10.41
+150 bps	7.73
+200 bps	5.13
+250 bps	2.60
+300 bps	0.15
+350 bps	−2.24
+400 bps	−4.55
+450 bps	−6.80
+500 bps	−8.99
+550 bps	−11.12

into a spot investment in a 5-year bond yielding 0.50% and a forward starting investment in a 5-year bond starting 5 years from today yielding 3.02%. The implied 5-year forward yield suggests an investor that has exposure to this steep part of the term structure (125 basis points from 5 year to 10 year) will earn a positive return on their investment even if 5-year yields increase by 250 basis points over the course of the next 5 years. Table 4.5 shows the expected return on this investment* under various interest rate scenarios.

The scenario analysis summarized in Table 4.5 shows the expected holding period return of an investment today in a zero-coupon, 10-year bond yielding 1.75% over the course of the next 5 years. In 5 years, the 10-year zero-coupon bond will age and effectively become a 5-year zero-coupon bond. If 5 years from now 5-year Treasury yields remained unchanged at 0.50%, today's investment in a 10-year zero-coupon bond would have gained 16.10% over the course of 5 years. In order for this investment to detract value, 5-year Treasury yields would have to yield over 3.50% (a 300 basis point increase from the 5-year spot level today). For example, if 5-year Treasury yields increased 550 basis points to yield 6.00% by 2017, the forward starting bond investment would have only declined by 11.12% over the course of 5 years. This holding period loss translates into an average annualized loss contribution of 233 basis points a year.

4.4.2 Risk Parity Is Not a Levered Bond Portfolio

Of course, there are periods in which bond yields would rise above the levels indicated by forward rates and bond exposures in a risk parity portfolio would lose value. However, this by no means would indicate the portfolio would lose value. This is because risk parity is not a levered bond portfolio and other parts of the portfolio: equity and inflation exposures are likely to have positive returns. This is precisely what happened in the episodes listed in Table 4.3.

* In this scenario the investment represents an initial investment in a 10-year zero-coupon bond today and measures the return on that investment 5 years later under various assumptions for the level of 5-year spot yields.

TABLE 4.6 Sharpe Ratio of Stocks, Bonds, and Risk Parity Portfolio and Correlation Between Stocks and Bonds

	2008	2009	2010
US WGBI	1.87	−0.69	1.37
S&P 500	−1.89	1.04	0.65
160/40 RP	0.09	0.17	2.99
Corr. US WGBI/S&P 500	−0.20	0.42	−0.73

The fundamental reason for this diversification is the role that different asset classes are expected to play throughout various macroeconomic environments or market cycles. As we have discussed in the previous sections, the performance of various forms of market risk premium is time varying and they are contemporaneously influenced by the macroeconomic environment or business cycle. For example, in a low-growth and low-inflation environment (similar to what we have experienced over the last several years), we would expect that that nominal sovereign bonds would deliver above-average performance. In contrast, in a high-growth environment, we would expect that equities and commodities would deliver above-average performance. A well-constructed risk parity portfolio balances the contribution to risk (and consequently contribution to return) across a diverse set of asset classes so that regardless of the environment neither the above-average nor the below-average performing asset classes will exert an undue influence on the total portfolio's performance. Risk parity portfolios achieve this balance by targeting balanced risk exposures. From this perspective, a risk parity portfolio's total performance as measured by its Sharpe ratio is related to the risk-weighted average of each asset class' component Sharpe ratios.

Table 4.6 summarizes the performance of a hypothetical risk parity portfolio consisting of a 160% notional weighting to the US World Government Bond Index and a 40% notional weighting to the S&P 500 Index. While the hypothetical risk parity portfolio delivered similar performance in 2008 (declining interest rates and equity prices) and 2009 (rising interest rates and equity prices), its best performance over the period was in 2010 when both equities and bonds delivered above-average risk-adjusted returns. Focusing on an asset's risk-weighted contribution to the total portfolio's Sharpe ratio rather than its notional weighted contribution is an important distinction that is often lost by risk parity skeptics. This oversight is the primary reason many people mistakenly believe that risk parity portfolios are levered bond portfolios excessively vulnerable to rising interest rate environments. Table 4.6 helps to dispel this myth. Just like in 2008 when poor equity market performance did not dominate the total portfolio, the poor bond market performance in the rising interest rate environment of 2009 did not cause the risk parity portfolio to endure a catastrophic drawdown.

4.4.3 Dynamic Risk Allocation

While bonds can contribute positively to portfolio performance even in a rising interest rate environment, it is dependent upon the steepness of the yield curve, the length of the cycle, and the magnitude of the yield increase. A scenario with a flat or inverted yield curve where interest rates spike higher over a short horizon would be a scenario where the allocation to nominal fixed income would significantly detract value from a risk parity portfolio. This scenario represents a structural repricing of market risk premium as the bond market was not expecting changes in interest rates (as evidenced by a flat yield curve) yet interest rates suddenly spiked higher. The last time this happened was in 1994.

In our view, structural market shifts require flexibility allowing the portfolio to become adaptive to changing market conditions. Therefore, it is desirable to have the flexibility afforded by a dynamic risk allocation process, which allows one to tactically shift the risk budget away from its long-term strategic targets in changing market conditions. For example, in an environment with a flat yield curve and spiking interest rates we would expect the dynamic risk allocation to bias us toward holding a lower risk allocation to fixed income from both a valuation perspective as well as a technical perspective. The flatness of the yield curve would suggest lower-term premium causing it to look less attractive from a valuation perspective.

4.4.4 Conclusion

A major concern of risk parity investors and prospective investors is how risk parity will perform in a rising interest rate environment. Despite embracing the intuitive appeal of constructing a risk-balanced portfolio, many investors struggle to get comfortable with the portfolio's unconventionally large notional exposure to fixed income in an environment with historically low fixed-income yield levels that are more likely to move higher than lower. In this note, we address this concern directly by making three points. First, the return contribution from fixed income is not guaranteed to be negative in a rising interest rate environment. When a yield curve is steep, and increases in interest rates occur over time, the portfolio's fixed-income positions can contribute positively to return if rates increase less than what is priced in by the forward yields. Second, a well-constructed risk parity portfolio should not be unduly influenced by the performance of any one particular asset class. In a rising interest rate environment, the magnitude of the loss contribution from the portfolio's fixed-income positions is not expected to be larger than the return contribution from other, uncorrelated assets (e.g., equities and commodities). Our experience managing risk parity portfolios for clients supports this expectation. In four out of the five periods with rising interest rates, the return contribution from equities and commodities more than offset the loss contribution from fixed income resulting in positive total portfolio returns. Finally, when interest rate increases occur as a result of a structural repricing in market risk premium it is important to have flexibility incorporated in your investment process to make the portfolio adaptive to changing market conditions. We suggest a dynamic risk allocation process to be systematically tactical across asset classes.

4.5 NO MORE RISK PARITY DEBATE?*

4.5.1 No Debate

Recently, a sell-side analyst invited me to be on a panel to debate the merits of risk parity. As either an enticement or a challenge, he said that he might be able to get someone

* Originally written by the author in April 2014.

from GMO to join me on the panel. I recall having mixed feelings about the prospect of debating someone like Mr. Ben Inker. On the one hand, Mr. Inker is a widely respected investor and researcher from a well-known firm, which has a sizable business built around traditional asset allocation investing. It would be indeed refreshing to have a thoughtful debate on risk parity with such a highly regarded investor. I would envision the discussion transcending beyond the typical fear mongering about the demise of risk parity when bond yields go up.

However, I am not sure we would find enough common ground to advance the debate. I view risk parity as an investment process built around true risk-based diversification that captures market risk premiums. In contrast, GMO's arguments against risk parity, while more articulate than the typical fear mongering, and presented with a strong value bent (Inker 2010, 2011), are built around a return-based forecasting framework. This perspective emphasizes return forecasts of individual assets that are quite uncertain rather than diversification, which is the key principle of risk parity. In other words, these critics miss the forests for the trees (or for the timber?).

Still, would it be thrilling to debate and maybe, in an extremely unlikely event, convert perhaps the most famous critic of risk parity? But of course it was not to be. Oh well, the sell-side analysts are not just optimistic on earnings estimates.

4.5.2 GMO 7-Year Return Forecasts

GMO has been arguing against risk parity, in the press at least, with papers and editorials since 2010. As recently as December 2013, it branded risk parity and other alternative or smart beta strategies together as snake oil (Montier 2013). After such a condemnation, it is impossible to know why they declined to debate this time. I have to admit that face-to-face debate is not exactly the most effective forum for the exchange of investment ideas. It is probably too superficial for most investment professionals, who would prefer the quiet format of expressing their thoughts in written words. In fact, maybe I should not agree to a debate either.

However, if there were ever a debate with someone from GMO, there is a very good counterargument to its continuing criticism of risk parity. It is generously supplied by GMO's own 7-year asset return forecasts, as of December 2013. If these forecasts were to be believed, a risk parity asset allocation portfolio would outperform a traditional 60/40 portfolio, at least for the next 7 years, of course.

Let us examine Figure 4.13, which displays the aforementioned return forecasts. The numbers are quite bearish for stocks. They are negative for US large cap and small cap stocks. Factoring in the real cash return of −0.4%, both US large cap and small cap stocks would have negative excess returns. Surprisingly or maybe not, GMO, always a proponent of quality stocks, expects US high-quality stocks to have a positive real return of 2.1%. The returns for bonds, while also low, are not as bearish. They are positive except for the dollar-hedged international bonds. The excess returns over cash are even more attractive (both on a relative and risk-adjusted basis). Besides timber, the highest forecasts are from the emerging markets, with emerging equities expected to return 3.5% and emerging debt expected to return 2.9%. It is not hard to deduce from these forecasts that stocks are considered quite overvalued and bond yields are assumed to rise modestly over time.

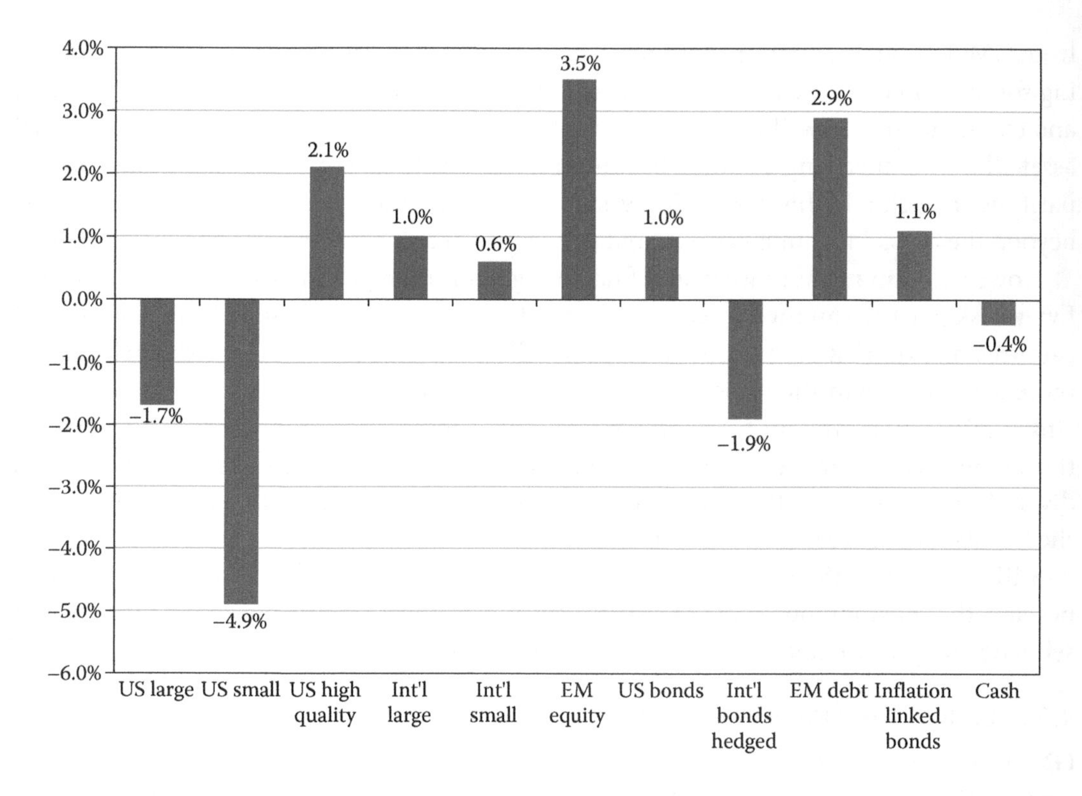

FIGURE 4.13 GMO 7-year real return forecasts at 12/31/2013.

Based on these return forecasts, I shall compare the expected returns of a risk parity portfolio and a traditional 60/40 portfolio, both hypothetic. The results show that the former has a higher expected return than the latter.

It is important to state from the outset that, this analysis in no way would suggest GMO might have or should have changed its view regarding risk parity, or may even have started its own risk parity fund, based on its own forecasts. Nor does it imply that the validity of risk parity shall hinge on this particular set of return forecasts, which might turn out to be right or widely off the mark. The analysis merely poses a simple question, could these forecasts be consistent with a risk parity approach?

4.5.3 Portfolio Expected Returns

Table 4.7 shows the expected asset class returns and weights of a 60/40 and a risk parity portfolio. We have omitted high-quality stocks, international small cap stocks, and timber, since most asset allocation portfolios do not have a dedicated allocation to them. Risk parity portfolios typically invest in commodities as a way to hedge inflation. However, it is even harder for anyone to forecast long-term commodity returns (e.g., Chapter 2), as former Fed Chairman Bernarke admitted when asked about the prospect of gold. In any case, timber does not represent commodities in general.

Traditional 60/40 portfolios are often capitalization-weighted while risk parity portfolios are risk-weighted. Under this general guideline, we put together two portfolios in

TABLE 4.7 Select Asset Classes and Portfolio Weights for a 60/40 and a Risk Parity Portfolio

	Real Return (%)	"60/40" (%)	"Risk Parity" (%)
US large	−1.7	25	15
US small	−4.9	5	10
Int'l large	1.0	25	15
EM stocks	3.5	5	10
US bonds	1.0	15	45
Int'l bonds hedged	−1.9	15	45
EM bonds	2.9	5	15
Inflation linked bonds	1.1	5	45
Cash	−0.4	0	−100

Table 4.7. The strategic capital allocation of the 60/40 portfolio has 25% each in both US large cap and international large cap, 5% each in both US small cap and EM stocks, for a total 60% in stocks. The 40% bond allocation is invested with 15% each in both US bonds and international bonds, and 5% each in both EM debt and inflation linked bonds.

For the risk parity portfolio, we have chosen a leverage of 200%, which comprises a capital allocation with 50% in stocks and 150% in bonds, respectively. The 50% stock investment comes from 15% each in both US large cap and international large cap, and 10% each in both US small cap and EM stocks. As for bonds, we assume an allocation of 45% each in US bonds, international bonds, and inflation-linked bonds, with the remaining 15% in EM debt due to its higher volatility.

Combining the expected returns and portfolio weights in Table 4.7 yields the expected portfolio returns shown in Table 4.8. The strategic 60/40 portfolio would deliver a real return of −0.18% while the strategic risk parity portfolio would deliver a real return of 0.68%. While both returns are dismal compared to historical averages because of the low-asset return forecasts, the risk parity portfolio does outpace the 60/40 portfolio by 0.86% per year.

4.5.4 Additional Advantage of Risk Parity

I would not expect critics of risk parity to surrender on the basis of 86 basis points. In fact, most traditional TAA managers argue that asset allocation portfolios should be enhanced through active management guided by the manager's ability to forecast asset returns. I agree with that to some extent. In fact, risk parity can also make tactical shifts away from the portfolio's strategic allocation to achieve desirable returns while also controlling portfolio risk, which is nearly impossible under traditional asset allocation approach. If the GMO forecasts are to be believed, as the following example illustrates, a risk parity

TABLE 4.8 Expected Portfolio Returns

	"60/40"	"Risk Parity"
Expected real return	−0.18%	0.68%

approach with active risk allocation offers a superior risk/return profile than any traditional long-only portfolio ever could.

Suppose an investor wants to achieve a targeted annualized return of 3.5% for the next 7 years. Purely based on a traditional asset allocation approach, the investor would have to put everything in emerging market equities, which is a highly risky move. Only an investor who had severe overconfidence in the precision and accuracy of their forecasts would be crazy enough to do that.

Using a risk parity approach, one can diversify portfolio risk with an appropriate use of leverage. For example, a portfolio of 35% in emerging market equities, 150% in US bonds, and 15% in emerging market debt, would achieve an expected return of 3.56%, with a much lower risk and more importantly far less sensitivity to global macroeconomic shocks. Compared to a portfolio with 100% EM equities, this risk parity portfolio is a more prudent, conservative portfolio for investors who are not averse to modest leverage but are rightly concerned about risk concentration and uncertainty in forecasts. In fact, because equities are inherently levered investments due to corporate borrowing, this risk parity portfolio probably has a lower effective leverage ratio than the portfolio with 100% EM equities.

4.5.5 Have the Critics Been Right about Risk Parity So Far?

One potential rebuttal to the previous analysis is "Sure, risk parity might work better given these low future expected equity returns. But the critics must have been right for the past few years when stocks have done so much better than bonds." Has risk parity done poorly versus 60/40 since Mr. Ben Inker first penned his criticism about risk parity?

To answer this question, we provide Figure 4.14 in which we plot the cumulative returns of the two representative portfolios from Table 4.7 since January 2010. It shows that since

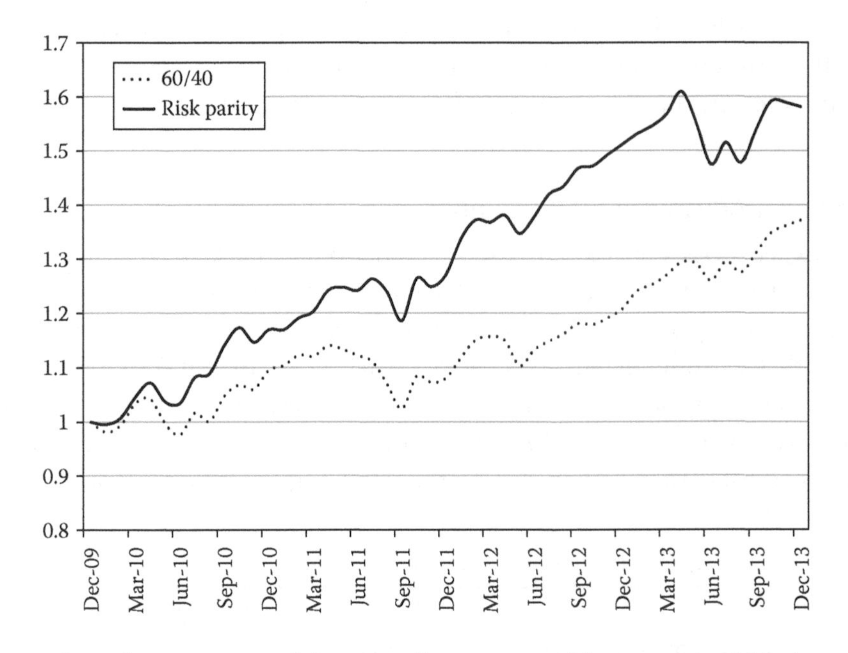

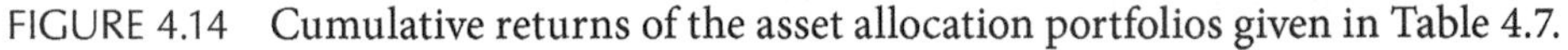
FIGURE 4.14 Cumulative returns of the asset allocation portfolios given in Table 4.7.

then, the risk parity portfolio has outperformed the traditional 60/40 portfolio by a sizeable margin. Yes, risk parity performed poorly relative to 60/40 in 2013 when the bond market repriced in May and June the future trajectory of Fed policy. However, the second quarter of 2013 was an exception rather than the rule. Over this 4-year period, the risk parity portfolio has an annualized return of 12.14% while the 60/40 portfolio has an annualized return of 8.23%. In addition, the two portfolios had similar annualized standard deviations. In other words, the risk parity portfolio has achieved one of its main investment objectives: a higher risk-adjusted return.

4.5.6 Conclusion

It seems both ex post returns and ex ante forecasts by GMO point to the superiority of risk parity over the traditional asset allocation approach. This would have been a concluding remark for this fictitious debate. So, no more risk parity debate? I hardly think so. "Everyone knows bond yields are going up!" I hear that uttering ringing again in the distance.

CHAPTER 5

The "Peculiarity" of Risk Parity Portfolios

Some of the questions regarding risk parity are about its portfolio implementation and management. Risk parity portfolios share many commonalities with traditional asset allocation portfolios but they are also different, and peculiar in several important aspects. In this chapter, we address some of these unorthodoxies related to portfolio leverage, portfolio rebalancing, benchmarking, and participation ratios.

First, the commonalities. Since risk parity multi-asset portfolios in multi-asset space are still asset allocation portfolios, their implementation in some way mirrors that of traditional asset allocation portfolios. For example, given notional asset weights of a risk parity portfolio that are derived based on risk allocation, one must choose ways to implement the asset exposures. In traditional asset allocation portfolios, one can use a combination of "passive" investments with index funds and "active" investments with actively managed strategies. For "passive" investments, one can choose between physicals or cash investments and derivatives such as index futures, government bond futures, and commodity futures.

The same is true for risk parity portfolios. For any asset exposure, one has the choice between passive and active, and between physicals and derivatives. However, there is one crucial difference: that is portfolio leverage. Since risk parity portfolios usually have leverage, one cannot implement all notional weights with cash investments. Remember capital is always 100% and notional exposure with leverage is greater than 100%. Therefore, some derivatives are required for risk parity implementation. However, there is no reason to require risk parity implementation to be all in derivatives. This is especially true if there are better ways to capture risk premiums that are not available with derivatives and thus require cash-based investments. In a risk parity portfolio, the efficient use of cash must be divided between physical investments for capturing risk premiums and cash as collaterals for derivatives.

We illustrate this point by using a risk parity portfolio with the following notional exposures in three-asset classes. Later in the chapter, we propose that this portfolio could serve as a simple benchmark for risk parity. As shown in Table 5.1, the portfolio has a leverage of 200%, stemming from 25% in commodities, 33% in equities, and 142% in government bonds. To implement this portfolio with a complete and thorough risk parity approach, we would build underlying asset exposures with risk parity commodity, risk parity equity, and risk parity fixed-income portfolios. For commodities, this should be achieved with all commodity futures. For stocks, it is not possible with futures, which are mostly based on country equity indices. Risk parity equity portfolios seek to balance risk across sectors as well as in individual stocks (more on this in Chapter 8). Hence, we would build risk parity equity with a portfolio of individual stocks. For fixed income, government bond futures exist for some countries (US, UK, Japan, etc.) but not for others. Here, we could use a combination of physicals, futures, and perhaps swaps. The point is by using futures and physicals judiciously, we would be able to implement a total portfolio with risk parity approach in both top-down and bottom-up.

Portfolio leverage also causes another difference between traditional long-only and risk parity portfolios in terms of portfolio rebalancing. Rebalancing traditional unlevered portfolios entails buying losers and selling winners. It is commonly known that this practice generates positive diversification return. Diversification return is not an excess return. However, when the return differences among different assets are small, portfolio rebalancing generates a positive rebalancing alpha. Things are somewhat different for levered portfolios. First, the leverage of portfolios changes along with profits and losses. When there is a gain, leverage declines and when there is a loss leverage increases. Hence, portfolio rebalancing on the top level entails buying (or relevering) after a gain and selling (or deleveraging) after a loss. Sometimes, this leads to buying winners and selling losers. The question is, does this lead to a destruction of diversification return?

In this chapter, we present four investment insights related to the topics mentioned above. The first insight is to demystify the role of leverage in risk parity portfolios. We argue that leverage is an essential tool of the modern economy and consequently a common presence in investors' portfolios. The second insight provides in-depth quantitative analysis of the effects of portfolio leverage on portfolio rebalancing and diversification return. Another frequently asked question related to portfolio implementation and performance monitoring is about potential benchmarks for risk parity portfolios. In the third sight, we list the pros and cons of some obvious choices as a benchmark for risk parity, ranging from cash plus, Sharpe ratio, to 60/40 portfolios. We go on to propose a simple risk parity benchmark based on equal risk allocation to three primary asset classes.

Some often view risk parity portfolios as a pure defensive investment strategy because of the exposures to lower volatility asset classes. This is only partly correct. In the last investment sight, we study the participation ratio of risk parity portfolios versus traditional

TABLE 5.1 Notional Weights of a Risk Parity Portfolio

	GSCI	MSCI	WGBI	Portfolio
Weight	25%	33%	142%	200%
Instrument	Futures	Physical	Futures and Physicals	

indices and benchmarks. It is true that risk parity, like a defensive strategy, has a low downside participation ratio. However, unlike a defensive strategy, its upside participation ratio is often close to one. This is another peculiar but desirable feature of risk parity portfolios.

5.1 WHO IS AFRAID OF LEVERAGE?*

Leverage became somewhat of a "dirty" word after the 2008 financial crisis. This is quite understandable. The crisis was partially caused by excessive leverage in the US housing market, the global financial markets, and by US consumers. Many investors became victims of the crisis because their portfolios were impaired by the horrific losses experienced by risky assets, which have only recently recovered. Now, in the aftermath of the crisis, deleveraging and austerity seem to be the prevailing macroeconomic themes. In a world of deleveraging, it would only be reasonable to think that everyone is afraid of leverage. Once bitten twice shy. No sooner had we started our risk parity strategy several years ago than the question of portfolio leverage was raised by consultants, asset managers, and prospects.

This purported fear of leverage in risk parity is not only irrational but also misplaced. Leverage, an indispensable tool in the modern economy and in investing, is neither inherently good nor bad. There are different kinds of leverage and it is important to make a distinction between them. In this research note, I demonstrate that the leverage embedded in a well-constructed risk parity portfolio is used properly to reduce portfolio risk concentration. In contrast, leverage embedded in, either explicitly or implicitly, traditional asset allocation portfolios often serves to increase portfolio risk concentration and it is the kind of leverage that investors should be leery of.

5.1.1 Modern Economy and Leverage

Before addressing the concern of portfolio leverage, we make some observations on the use of leverage in a modern economy. Although we usually do not think about it, leverage is a part of life in a modern society.

First, modern economy depends on credit to function. One way to create credit is through fractional-reserve banking. Since commercial banks are only required to keep a fraction of deposits on reserve, they lend out the rest as loans. Some of these loans subsequently are deposited in another bank, leading to another round of lending, net of the fractional reserve on the new deposits. Therefore, the fractional-reserve system generates money supply (currencies plus reserves plus deposits) beyond the base money that is created by the central bank, by increasing the aggregate amount of deposits. This multiplier effect creates leverage, which has a maximum theoretical value of $1/r$, where r is the reserve requirement. For example, if the reserve requirement is 10%, then the maximum multiplier is 10. One of the tools used by a central bank is to change a reserve requirement. Although no longer a conventional tool in developed countries, it is one of the major policy levers in China.

Figure 5.1 shows the ratio of the money supply measured by M2 to the monetary base in the United States. Prior to 2008, it fluctuated around 8 but it collapsed to near 4 as the Fed started various asset purchasing programs near the end of 2008. As the excess

* Originally written by the author in February 2012.

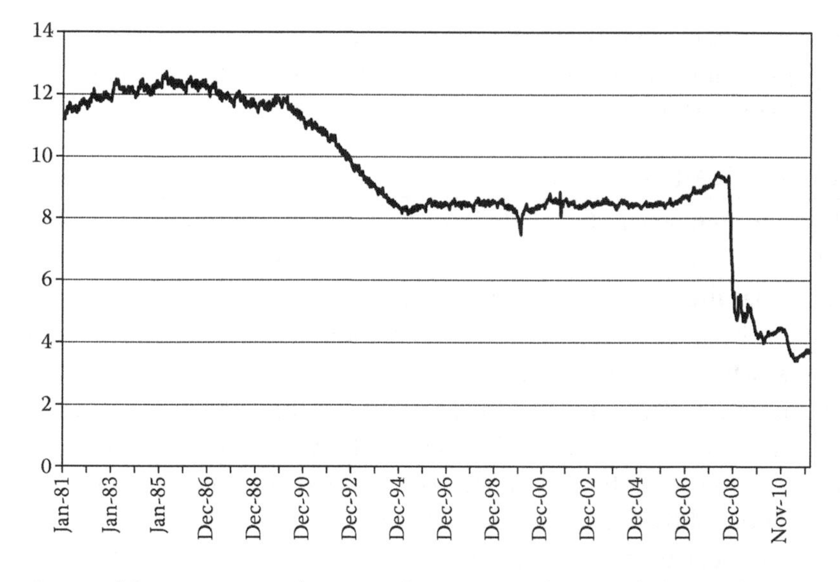

FIGURE 5.1 Ratio of the money supply M2 to the monetary base in the United States.

reserves created by asset purchasing programs (QE) increased dramatically, they sat idle on the Fed's balance sheet rather than being lent out by commercial banks. As a result, the increase in the money supply has not kept pace. It is a sign of deleveraging even as the Fed is trying its best to leverage its own balance sheet.

Another form of leverage in a modern economy is corporate debt. Corporations borrow from banks and debt markets to fund ongoing operations and to expand their businesses. This is, of course, obvious for financial institutions, which owe their very existence to leverage. However, even for nonfinancial corporations, the leverage is quite substantial. Figure 5.2 shows the ratio of total liabilities to total net worth of nonfarm and nonfinancial corporations in the United States. The ratio reached the highest level during the 1990 recession, and it has since declined. The ratio increased sharply during the 2008 crisis but resumed its descent after the recession ended. It appears that the US corporations have been repairing their balance sheets since the 1990s and they continue to deleverage after the 2008 crisis.

Finally, the leverage of the household sector is also a big factor in the US economy. In a consumption-driven economy, consumers get loans for homes, cars, or get credit lines (e.g., credit cards) for other expenses. Figure 5.3 shows the ratio of household debt to disposable income over time. The overleveraging of US consumers caused the ratio to double from the 1980s' to 2007 and has since declined from the peak of 135% to 120%. It appears that the deleveraging still has some ways to go.

In summary, leverage is an inescapable part of modern society and modern economy. Thus, it should be recognized that properly employed leverage can support and accelerate economic growth. An example of perhaps the most successful use of leverage is Warren Buffet's Berkshire Hathaway, which uses insurance premiums to invest in the stock market and directly in real businesses. Neither direct borrowing nor financial leverage is involved, but it is leverage nevertheless with insurance liability. On the other hand, excessive leverage

FIGURE 5.2 Ratio of total liability to total net worth of nonfarm and nonfinancial corporate businesses in the United States.

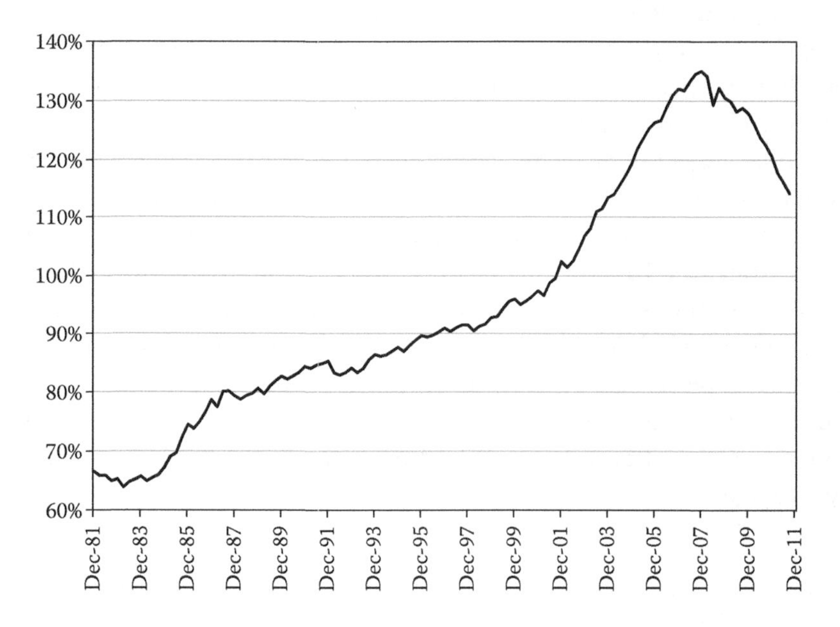

FIGURE 5.3 Ratio of household debt to disposable income.

created by shadow banking, over-indebted corporations, and debt-financing consumption can lead to economic hardships.

5.1.2 Institutional Investors and Leverage

It must be noted that institutional investors have a similar affinity for portfolio leverage as the broader economy even though some may not be aware of it. To start with, a 60/40 portfolio would have an implied leverage ratio 160% under the assumption that public equity is levered 2:1, which is consistent with evidence in Figure 5.2. Over the last decade, however, institutional portfolios of endowments and foundations, and pension funds have embraced a shift from traditional 60/40 portfolios to portfolios with many alternative asset classes that employ varying degrees of explicit financial leverage. They include private equity (or leveraged buyouts to be more precise) and real estate investments, both of which are levered using loans. In addition, many institutional portfolios maintain a healthy exposure to hedge funds, which are levered using both long and short positions.

In Table 5.2, we show a hypothetical institutional portfolio with six asset classes: two traditional asset classes fixed income and public equity, and four alternative asset classes: private equity, real estate, hedge funds, and commodities. We assume fixed income and commodities are fully funded or collateralized. We assume the leverage ratio for public equities is 2:1. We also assume a 2:1 leverage ratio for hedge funds, which is consistent with a market neutral long-short equity hedge fund. This is a fairly conservative assumption as GM hedge funds and other types of hedge fund strategies tend to have much higher leverage. For private equity and real estate, it is hard to find public data on the degree of leverage used. In addition, it is also the case that the leverage ratio could differ drastically across different funds. Regardless, they seem to have higher leverage than public firms have. We assume a leverage ratio of 3:1 for private equity and 5:1 for real estate. Table 5.2 applies the various asset class leverage ratios to a typical asset class weighting scheme to estimate the leverage of a hypothetical institutional portfolio. Under these assumptions, the portfolio has an aggregated leverage ratio of 235%.

This level of leverage might seem surprising and even shocking. Perhaps it should not be, for a couple of reasons. First, most of the leverage is implicit, either because it is hidden inside the investments such as public equity or because they are taken up by managers of underlying strategies. Investments in these strategies are done through a fund structure, which renders the leverage out of sight. The point is whether leverage is apparent or not on the portfolio balance sheet, it is real and has enormous economic and financial consequences. As it pertains to leverage in institutional portfolios, out of sight should not be out of mind. Second, we live in a broadly leveraged economy. The implication of a leveraged economy is that it is difficult to find many investments that are "pure" and not affected by leverage.

TABLE 5.2 A Hypothetical Portfolio with Six Asset Classes and Assumed Leverage Ratios

	Weight (%)	Leverage Ratio	Leveraged Weight (%)
Fixed income	15	1:1	15
Public equity	30	2:1	60
Private equity	15	3:1	45
Real estate	15	5:1	75
Hedge fund	15	2:1	30
Commodities	10	1:1	10
Total	100	2.35:1	235

5.1.3 Danger of Hidden Leverage

Of course, the fact that leverage is ubiquitous should not lull us to think leverage is always safe. For one thing, excessive leverage, especially those applied to illiquid investments is always extremely dangerous because it can result in catastrophic losses over a short time horizon. We only have to look at the cases of Bear Sterns, Lehman Brothers, or AIG to know that a leverage ratio of 30:1 or more could be troublesome. In the case of AIG, one can argue that the leverage was infinite if the firm did not put any collateral aside for insuring subprime mortgages. LTCM provides an investment example in which the portfolio leverage ratio was believed to be in the hundreds.

Another kind of portfolio leverage, which might seem modest but could still be problematic, is the leverage that exacerbates a portfolio's risk exposure to a single risk dimension. I believe the leverage embedded in the hypothetical institutional portfolio in Table 5.2 or in the traditional 60/40 portfolio belongs in this category. For example, in the 60/40 portfolio, the equity risk exposure, which is mostly risk associated with economic growth, typically represents 95% of a portfolio's total risk. The institutional portfolio suffers the same imbalance in risk exposure since four of the six asset classes, public equity, private equity, real estate, and commodities, are highly exposed to growth risk. Even hedge funds, which are supposed to be market neutral, usually contain a high degree of market exposure. From the perspective of growth risk exposure, the institutional portfolio and a 60/40 portfolio have more in common than what their asset allocations would imply. This is confirmed by their similar performance during the 2008 financial crisis and during the subsequent recovery of risky assets.

5.1.4 Risk Parity and Leverage

Risk parity's leverage is explicit but different. The purpose is not to increase portfolio risk or to tilt the portfolio in a particular risk dimension to increase returns. We use leverage to balance portfolio risk exposures across asset classes. As a result, well-designed risk parity portfolios are truly diversified and expected to deliver a higher Sharpe ratio than the traditional 60/40 portfolio.

For example, assume stocks and bonds have volatilities of 20% and 5%, respectively. A risk parity portfolio with a similar total risk to that of a 60/40 portfolio might result in a 40/160 portfolio. If we include the implicit leverage of equity, then the total leverage would be 240%, which is similar to the embedded leverage of the institutional portfolio.

However, in terms of risk exposure, risk parity is balanced between growth risk and recessionary risk, as it captures both equity and interest rate premiums, while also providing downside protections during financial stress. However, the 60/40 or the institutional portfolio is heavily exposed to growth risk. This is why we think that the leverage in risk parity is used prudently.

5.1.5 Conclusion

During the 2008 financial crisis, risky assets that had either implicit or explicit leverage to growth risk suffered significant losses. Since the crisis was brought on by the excessive leverage of homeowners and financial institutions, the use of portfolio leverage has been questioned by many investors.

It is important to maintain a proper perspective regarding different kinds of leverage including, but not limited to, implicit versus explicit, modest versus extreme, and the kind that creates risk concentration versus the kind that diversifies risk. Upon closer examination, many institutional investors invest in assets and strategies that have implicit leverage and create risk concentration. On the other hand, risk parity uses leverage explicitly in a way that diversifies risk. In the end, it might be the case that investors should be more concerned about the leverage in a traditional 60/40 portfolio and other similar risk-concentrated portfolios rather than the leverage in risk parity portfolios.

5.2 PORTFOLIO REBALANCING AND DIVERSIFICATION RETURNS OF LEVERAGED PORTFOLIOS*

As 2011 draws to a close, I find myself pondering a simple investment idea—portfolio rebalancing. Before acting on the so-called top 10 investment ideas of 2012, the year end is a good time for investors to consider rebalancing portfolios to their prescribed weights. Why is portfolio rebalancing so important? It is the simplest technique that could add incremental value to fixed-weighted multi-asset class portfolios. Part of this incremental value is often referred to as diversification return.†

I like to use the following example to illustrate the power of diversification and portfolio rebalancing. Suppose we have at our disposal two investments. Investment A doubles in the first year and then promptly drops by half in the second year. In contrast, investment B moves in the opposite way: it goes down by 50% in year 1 and then recovers by 100% in year 2. After 2 tumultuous years, both investments have gone nowhere individually, However, a 50/50 portfolio with 50% of its capital in each would yield a return of 25% in both year 1 and year 2 (rebalancing!) without annual return volatility.

In this example, diversification and portfolio rebalancing rebalance go hand in hand and it is hard to separate their contributions. Without rebalancing, any portfolio, regardless of its initial weights, would not have any gain. Similarly, a concentrated portfolio (e.g., 100% in either investment A or B) would be devoid of any rebalancing opportunity and it would yield zero return. Among all possible portfolios of the two assets, the 50/50 portfolio is the most diversified in terms of generating the highest diversification return through rebalance.

Portfolio rebalancing is a well-known technique and an old research topic. For instance, numerous articles have been written regarding different rebalancing techniques: calendar rebalance, threshold rebalance, tactical rebalance, etc. So why should we spend more time on the topic? The reason is that all the previous research focuses almost exclusively on long-only portfolios without leverage. What would happen for portfolios like risk parity portfolios that are leveraged, or portfolios with both long and short positions? Does portfolio rebalancing still make sense? Does it still generate incremental value? These are important questions as investors continue to embrace the concept of risk parity and appreciate the

* Originally written by the author in December 2011.

† Diversification return is only part of rebalancing alpha. Another part is the return effect from the dispersion of returns, which is typically negative for long-only portfolios. The net effect of these two parts is rebalancing return. It is true that in theory, rebalancing alpha should not exist if asset returns are serially independent. However, in practice, many asset returns exhibit mean reversion over various time horizons, leading to positive rebalancing alpha.

benefits of appropriately used leverage. In this research note, I shall address these questions and share some insights.

5.2.1 Rebalancing Long-Only Unlevered Portfolios

Before considering leveraged portfolios, let us review some results concerning fixed-weighted long-only unlevered portfolios. For a calendar-based rebalancing scheme, the exercise is simple: periodically a portfolio is rebalanced back to its original weights. However, to understand it further and extend our understanding to levered portfolios, we note that the rebalance is carried out as a contrarian or mean-reverting strategy.

In our simple example, the 50/50 portfolio drifts to an 80/20 portfolio (100/125 = 80%, 25/125 = 20%) after year 1. In order to rebalance the portfolio to the original 50/50 weight, we sell 30% of asset A (the winner) and buy 30% of asset B (the loser). This mean-reverting strategy is also true for portfolios with more than two assets. The weights of the assets that have positive excess returns versus the portfolio will drift higher while the weights of assets that have negative excess returns versus the portfolio will drift lower. As a result, rebalancing necessarily requires selling the former group (the winners) and buying the latter group (the losers).

Why would such a mean-reverting strategy generate positive value added versus a buy-and-hold strategy for a long-only portfolio? The reason is it takes advantage of randomness or volatility of asset returns, which is always present regardless of how different the average returns of the individual assets might be. In the simplest case, where all assets have identical cumulative return over time, the mean-reverting strategy would sell assets that happen to overshoot the long-term average and buy assets that happen to undershoot their long-term average. As a result, the mean-reverting strategy generates positive diversification return versus the weighted sum of the individual returns.

5.2.2 Diversification Return

It is time we formally define diversification return: it is the difference between the geometric return of a multi-asset portfolio and the weighted sum of the geometric returns of the underlying asset classes. It is crucial to see that the geometric average return, not arithmetic average return, is what matters. We now turn to mathematical analysis to gain insights about diversification return. We define the diversification return as the difference between the return of the fixed-weight portfolio and the weighted average of individual returns:

$$e_v = g_{FW} - (w_1 g_1 + \cdots + w_M g_M), \tag{5.1}$$

where g denotes the geometric return.

TABLE 5.3 Investment History of a 50/50 Portfolio with Rebalancing

	$t = 0$	$t = 1$ (Before)	$t = 1$ (After)	$t = 2$
Investment A	\$0.5	\$1($w_A = 80\%$)	\$0.625($w_A = 50\%$)	\$0.3125
Investment B	\$0.5	\$0.25 ($w_B = 20\%$)	\$0.625($w_B = 50\%$)	\$1.25
Portfolio total	\$1	\$1.25	\$1.25	\$1.5625

This diversification return is always nonnegative for fixed-weighted, long-only, unlevered portfolios. It suffices to state here that the key to the proof is that the variance of a fixed-weighted portfolio is in general smaller than the weighted sum of the individual variances because of diversification. This variance difference is approximately twice the diversification return defined above. As a result, the geometric return of the portfolio is greater than the weighted sum of the geometric returns of the individual assets. In the case in which all assets have zero cumulative return, as in my two-asset example, portfolio rebalancing can generate a positive return over time—in other words, the proverbial something out of nothing.

For clarity, we shall continue to use portfolios of two assets as illustrations throughout this research note. The following formulas show how diversification return is calculated for a two-asset portfolio. Assuming portfolio weights (w_1, w_2), volatilities (σ_1, σ_2), and correlation ρ_{12}, the diversification return is given by

$$\begin{aligned} e_v &= 0.5(w_1\sigma_1^2 + w_2\sigma_2^2 - \sigma_p^2), \\ \sigma_p^2 &= w_1^2\sigma_1^2 + w_2^2\sigma_2^2 + 2w_1w_2\rho_{12}\sigma_1\sigma_2. \end{aligned} \tag{5.2}$$

The first equation states diversification return is half of the difference between the weighted sum of the two individual variances and the portfolio variance. The second equation provides the portfolio variance in terms of the weights and the variances and the correlation. For long-only unlevered portfolios, we use this result to analyze a 60/40 portfolio with stocks and bonds. Assuming volatility of 20% and 5% for stocks and bonds, respectively, the diversification returns would be 0.63%, 0.51%, and 0.39% for three different correlations: $\rho = -0.5, 0, 0.5$. The results are summarized in Table 5.4, together with three portfolio volatilities.

Table 5.4 shows that lower correlation not only leads to lower portfolio volatility, but also a higher diversification return. There are important implications of this diversification return analysis to asset allocation decisions between stocks and bonds. It is known that the correlation between stocks and bonds varies greatly when the bonds are of different credit quality. The correlation is typically negative when the bonds are high-quality government bonds. The correlation is near zero if one invests in aggregate bond indices. The correlation is highly positive if one invests in corporate bonds and HY bonds. In the last case, not only is the benefit of risk reduction small but the high correlation also tends to decrease the diversification return.

5.2.3 Rebalancing Leveraged Portfolios

In contrast, rebalancing leveraged portfolios involves momentum or trend following, that is, buying winners and selling losers. While this might appear surprising and perhaps counterintuitive, a simple example suffices to illustrate the logic.

TABLE 5.4 Diversification Returns and Volatilities of 60/40 Portfolios

60/40	$\rho = -0.5$ (%)	$\rho = 0$ (%)	$\rho = 0.5$ (%)
Diversification return	0.63	0.51	0.39
Portfolio volatility	11.1	12.2	13.1

TABLE 5.5 Investment History of a Winning Long–Short Portfolio with Rebalancing

	$t = 0$	$t = 1$ (Before)	$t = 1$ (After)
Investment A	\$2	\$3($w_A = 150\%$)	\$4 ($w_A = 200\%$)
Investment B	\$–1	\$–1 ($w_B = -50\%$)	\$–2($w_B = -100\%$)
Portfolio total	\$1	\$2	\$2

Consider in Table 5.5 a two-asset 200/100 portfolio with \$2 (200% long) in asset A and –\$1 (100% short) in asset B. Suppose asset A returns 50% and asset B returns 0%. At the end of period, asset A grows to \$3 and asset B stays at –\$1 so the net value of the portfolio doubles. As a result, the portfolio weight shrinks to 150/50 (300/200 = 150% and 100/200 = 50%). To rebalance the portfolio to the original 200/100 target weights, we would buy 50% of asset A (the winner) and sell 50% of asset B (the loser). It is easy to prove mathematically that when a leveraged portfolio has positive returns, gross leverage declines and one would have to increase leverage to get back to the original weights.

The opposite is true when a leveraged portfolio suffers losses. In Table 5.6, suppose asset A returns –25% and asset B returns 0%. Asset A declines to \$1.5 and asset B stays at –\$1 so the net value of the portfolio is halved. As a result, the portfolio weight is now 300/200 (\$1.5/\$0.5 = 300% and \$1/\$0.5 = 200%). To rebalance the portfolio to the original 200/100 mix, we would sell 100% of asset A (the loser) and buy (or short covering) 100% of asset B (the winner). When a leveraged portfolio has negative returns, gross leverage increases and one would have to cut leverage or stop loss to get back to the target weights. In this context, the practice of stop-loss or deleveraging, which is sometimes hailed as a distinctive investment decision by hedge fund managers, at times when fresh capitals is not readily available, can be viewed merely as a mechanical decision to rebalance long-short portfolios.

So does this trend following rebalance strategy generate diversification return? If it does, will the diversification return become negative? The answer is that it is negative in some cases and positive in other cases.

5.2.4 Negative Diversification Returns: Cautionary Tales of Inverse and Leveraged ETFs

The simplest example of leveraged portfolios that have negative diversification returns are short portfolios or leveraged portfolios of a single risky asset. Nowadays, one does not have to look long for real-world examples. Over the last few years, there has been a proliferation of inverse and leveraged exchange-traded funds (ETFs). The former typically takes short positions in cap-weighted indices while the latter takes leveraged exposure to cap-weighted indices, with the aid of derivatives. The positions or weights are rebalanced daily, hence the

TABLE 5.6 Investment History of a Losing Long–Short Portfolio with Rebalancing

	$t = 0$	$t = 1$ (Before)	$t = 1$(After)
Investment A	\$2	\$1.5($w_A = 300\%$)	\$1($w_A = 200\%$)
Investment B	\$–1	\$ – 1 ($w_B = -200\%$)	\$ – 0.5($w_B = -100\%$)
Portfolio total	\$1	\$0.5	\$2

daily returns of these ETFs mirror the prescribed multiple of the underlying indices. But over the long run, they are most likely to lag the index multiples.

The daily rebalancing of these inverse and leveraged ETFs amounts to buying high and selling low. However, it is not widely recognized that the fundamental cause of the return slippage is the fact that portfolio rebalancing of these ETFs to constant leverage ratios generates a negative diversification return. For portfolios with a single risky asset and the risk-free asset (i.e., cash)—a necessary component for inverse or leveraged portfolios, the formula for the diversification return reduces to

$$e_v = 0.5(w_1 - w_1^2)\sigma_1^2, \tag{5.3}$$

where w_1 is the weight of the risky asset and σ_1 is the volatility of the excess return of the risky asset. For an inverse ETF, $w_1 = -100\%$. For an ultralong 2X ETF, $w_1 = 200\%$. It is easy to see both weights, when substituted into the equation, results in a negative diversification return. However, for long-only portfolios, in which $0 < w_1 < 100\%$, the diversification is always positive.

So how large is the potential return slippage—the difference between the return of the ETF and the multiple of the index return? We provide numerical examples by assuming the risk index has an annual volatility of 20% (the likes of the S&P 500 index) and a daily volatility of about 1.3% ($=20\%/\sqrt{250}$), assuming daily returns are independent. Using Equation 5.3 above we obtain the daily return slippage for the following five ETFs, –3X, –2X, –1X, 2X, and 3X, shown in the second row of Table 5.7 below. The next row displays the annualized slippage. The last two rows show the annual returns of the five ETFs when the underlying index return is –5% or 5% per year, respectively.* The annual slippage of –3X ETF is a whopping –21.3% while that of 3X ETF is –11.3%. The return slippage creates significant return hurdles for these leveraged and inverse (or both) ETFs. For example, when the index annual return is –5%, a naïve investor might expect the super bearish –3X ETF to yield +15%, but in reality, the likely outcome (gross of fees) is negative, which is

TABLE 5.7 Daily and Annualized Diversification Returns (slippage) of Inverse and Leveraged ETFs of an Index, Whose Annual Volatility is 20%

	–3X (%)	–2X (%)	–1X (%)	2X (%)	3X (%)
Daily	–0.10	–0.05	–0.02	0.00	–0.02
Annual	–21.3	–11.3	–3.9	–3.9	–11.3
Index –5%	–6.3	–1.3	1.1	–13.9	–26.3
Index 5%	–36.3	–21.3	–8.9	6.1	3.7

* The result in Equation 5.3 is an approximation that is only valid when asset returns are relatively small and leverage is moderate. Due to this limitation, results presented in this chapter are not intended for portfolios that are levered 20 or 30 to 1. In addition to excessive leverage, managers of these portfolios violate our rebalance assumption—they do not practice it rigorously. They probably do it to true up leverage in good times when their portfolios increase in value, that is, buying more risky assets on the way up. As recent experience shows, they certainly do not rebalance when their portfolios lose significant value, since in that case they either go under or get bailed out, by a parent company or taxpayers, rendering rebalance either unattainable or unnecessary.

−6.3% (=15%−21.3%). On the other hand, an investor in a 3X ETF might dream of a 15% annual return when the index returns 5% in a year. The likely outcome (gross of fees) is only 3.7% (=15%−11.3%), which is actually less than the return of the plain vanilla index itself!

In my view, while these inverse and leveraged ETFs based on a single risky asset might be an efficient tool for short-term investments, they are ill-suited as long-term investments due to the large negative diversification returns.*

5.2.5 Diversification Return of Leveraged Risk Parity Portfolios

We now turn to the diversification return of risk parity multi-asset portfolios. The key insight is that the rebalance of a long-only, fixed-weighted, leveraged portfolio with multiple risky asset classes combines two separate strategies on the two separate levels analyzed above. The first level is bottom-up, in which the mean-reverting strategy is carried out across the individual asset classes. The second level is top-down, in which the trend following strategy is employed on the total portfolio that is leveraged. As we demonstrated above, the bottom-up mean-reverting strategy generates positive diversification return, which is amplified by the leverage in this case, while the top-down trend following strategy gives rise to negative diversification return. The net result (Qian 2012), which can be either positive or negative, depends on the asset allocation mix as well as the overall leverage of the portfolio.

We use stock/bond portfolios as an example to illustrate the point. First, an unleveraged risk parity portfolio with stocks at 20% volatility and bonds at 5% volatility is a 20/80 portfolio. Table 5.8 shows the diversification returns and portfolio volatilities of the 20/80 portfolios with three different correlations. Notice the volatilities are quite low because only 20% is allocated to stocks. It is also noted that the level of diversification return is lower than that of the 60/40 portfolios. This is not too surprising since the maximum diversification return is achieved with the 50/50 equal-weighted portfolio. A 60/40 portfolio is "closer" to the 50/50 portfolio than a 20/80 portfolio.

Risk parity portfolios use leverage to balance risk contribution as well as to target total portfolio risk. One targeted risk is that of the traditional 60/40 portfolios. We thus choose leverage ratios on the 20/80 portfolio such that the portfolio risks match those of the 60/40 portfolios. Table 5.9 lists the leverage ratios for three different correlation assumptions and the resulting diversification return.

When the correlation is −0.5, the 20/80 portfolio risk is only 4% and the required leverage is relatively high at 278%. At 278% leverage the diversification return increases from

TABLE 5.8 Diversification Returns and Volatilities of 20/80 (Unleveraged Risk Parity) Portfolios

20/80	$\rho = -0.5$ (%)	$\rho = 0$ (%)	$\rho = 0.5$ (%)
Diversification return	0.42	0.34	0.26
Portfolio volatility	4.00	5.66	6.93

* For 2011, when the S&P 500 index returned a 2.1% the leveraged ETFs would have negative returns.

TABLE 5.9 Returns, Volatilities, and Leverage Ratios of Risk Parity Portfolios

Risk Parity	$\rho = -0.5$ (%)	$\rho = 0$ (%)	$\rho = 0.5$ (%)
Diversification return	0.77	0.34	0.08
Portfolio volatility	11.1	12.2	13.1
Portfolio leverage	278	215	190

0.42% to 0.77%, which is actually higher than that of 60/40, which is 0.63%. With negative correlation between the two assets, the bottom-up mean-reverting strategy dominates the top-down trend-following strategy. The net result is an increase in diversification even though the leverage ratio is rather high.*

When the correlation is zero, the required leverage declines to 215%. In this scenario, the diversification return stays at 0.34% and the bottom-up and the top-down effects caused by leverage are roughly equal. However, the diversification return of 0.34% is still lower than that of the 60/40 portfolio at 0.51%. When the correlation is 0.5, the required leverage is only 190%. In this case, the leverage lowers the diversification return from 0.26% to 0.08% since the high correlation between the two assets limits the diversification return from the bottom-up effect.

Comparing the three cases, one includes that low correlation between the two assets is key to maintaining a high level of diversification return as the underlying portfolio is being leveraged. From this perspective, high-quality government bonds again serves as a better diversifier than corporate or low-quality bonds because the latter has high correlation with stocks.

Another noticeable fact of Table 5.9 is that in all three cases the diversification returns of the leveraged portfolios remain positive despite leverage. Of course, this cannot be true if the leverage ratios are much higher especially in the case of positive correlation. Therefore, it is important that the leverage ratio be kept at levels that are consistent with a positive diversification return.

5.2.6 Conclusion

This research note provides some analytical results regarding portfolio rebalancing and the associated diversification returns for different kinds of portfolios. For long-only unlevered portfolios, we show that rebalancing amounts to mean-reverting strategies and the diversification return is always positive for multi-asset class portfolios. Therefore, adherence to regular portfolio rebalancing is strongly recommended.

For short (or inverse) and leveraged portfolios, we provide the key insight that rebalancing on the top-down level amounts to trend-following strategies that detract diversification returns while rebalance among individual assets at the bottom-up level is still mean reverting and adds to the diversification return.

We highlight the pitfalls of inverse and leveraged ETFs based on a single risky asset, for instance, a cap-weighted equity index, which often carries a significant negative diversification return, or return slippage over time, against index return multiples.

* In practice, risk parity portfolios use more than two asset classes and varying leverages but the same principle applies. We find our version of risk parity multiasset portfolio has a diversification return of approximately 0.75%.

Skeptics might argue that cap-weighted indices are not exactly a single risky asset. It is imperative to realize that even though cap-weighted indices might consist of hundreds or even thousands of securities, they are not fixed-weighted portfolios and hence offer no diversification return since cap-weighted indices never rebalance. In fact, this might be the Achilles' heel of cap-weighted indices, apart from their risk concentration in countries, sectors, or stocks. Without regular rebalance, frequent booms and busts of individual stocks, sectors, and countries within the indices simply take investors on wild rides that go nowhere over time, not too dissimilar in spirit to portfolios of investments A and B in our simple example that are not rebalanced. Viewed in this light, equal-weighted indices might seem to be naïve, but regular rebalancing at least provides a guaranteed positive diversification return over time. This crucial difference goes a long way to explain why equal-weighted indices often beat cap-weighted indices in the long run.

Our results concerning risk parity portfolios show that the diversification return is in general positive for leveraged risk parity portfolios when the leverage ratio is not too high (under 300%) for two-asset class portfolios with stocks and bonds. For portfolios with more asset classes, the leverage ratios can still be higher due to the fact that the bottom-up diversification effect will be higher with more assets.

Another important factor in determining portfolio diversification return is the correlation between different assets. The lower the correlation is, the higher the diversification return. Our stock/bond example strongly suggests that the best diversifying assets to stocks are high-quality government bonds due to its negative correlation to stocks. In contrast, the worst diversifying assets to stocks might be low-grade corporate bonds due to their relatively high correlation to stocks.

5.3 BENCHMARKING RISK PARITY*

Even risk parity investors who are disenchanted with traditional asset allocation approaches due to their lack of diversification cannot quite let go of the "good" old 60/40 portfolio. This is likely because a 60/40 stock/bond portfolio remains the most commonly used benchmark of asset allocation strategies. Risk parity portfolios follow an entirely different portfolio construction process. While they could be either passive or active in their own right, in order to achieve true risk diversification they should never be actively managed against a 60/40 portfolio. If this is the case, can the 60/40 portfolio serve as a benchmark?

Several problems could arise when there is a mismatch between portfolios and their benchmarks. The first is that return differences are often large, especially in the short term, violating the normal relationship between a portfolio and the benchmark to which investors are accustomed. A second but related problem is that even though it is expected that over the long run risk parity portfolios should outperform 60/40 portfolios, the information ratio (IR)—a measurement of active investment—can be quite low due to the large tracking error mentioned above. As a result, some investors may wrongly conclude that risk parity portfolios are not worth considering. Of course, this is caused by confusion between strategic and tactical asset allocation. However, the biggest problem with regard to

* Originally written by the author in January 2013.

an inappropriate benchmark is that it deprives investors of proper performance attribution that is crucial for an objective evaluation of the success and sustainability of an investment strategy.

There is perhaps a history lesson in the case of hedge funds. Hedge fund indices are riddled with selection bias and survivorship bias. The indices could be either equally weighted or asset weighted; however, neither is investable due to the illiquidity of individual funds and hedge funds in general. In addition, individual hedge funds are wildly different from each other and they often do not resemble the indices or even the subindices for specific strategies. As a result, investors in hedge funds do not have clear benchmarks that can be used in assessing manager's return sources, their skills or their luck.

In hindsight, the lack of proper benchmarks did not dampen either the acceptance or the perceived success of many alternative investment strategies. Perhaps, it should have.

So what is an appropriate benchmark for a risk parity strategy? Are the traditional 60/40 portfolios appropriate? In this research note, I will analyze possible choices for risk parity, including a cash benchmark, Sharpe ratio benchmark, and the traditional 60/40 portfolio. In addition, I shall propose a different, but more suitable, benchmark for risk parity portfolios.

5.3.1 Investment Objectives and Benchmarks

It is important to note that a benchmark for a given investment strategy must bear some relationship to its investment objective. For many active and passive strategies aimed to outperform or match specific indices, the relevant benchmarks in relative terms are those indices that the strategies are designed to outperform.

However, for many alternative investment strategies that aim to deliver absolute returns, the investment objective is not so clear-cut and this ambiguity often leads to multiple choices of benchmarks. First, to invest risk-free cash in any investment strategy is to take on risk, which should be rewarded in terms of excess return over cash. Therefore, one natural benchmark is cash, or more often than not, cash plus a defined percentage. Second, another investment objective of many alternative investment strategies, especially hedge fund strategies, is to deliver stable returns regardless of overall market conditions. This translates to risk-adjusted returns and therefore another benchmark would be Sharpe ratio or its variants, which are based on risk-adjusted excess returns. However, there are at least two complications with a Sharpe ratio benchmark. The first is defining an appropriate level of Sharpe ratio. The other issue is that a good Sharpe ratio does not necessarily imply a good return when the strategies' risk levels are not specified.

5.3.2 Serving Multiple Masters

What does our discussion mean for risk parity strategies? We start with their investment objective: providing stable returns by capturing various return premiums in a risk-balanced portfolio. This somewhat lengthy statement not only claims an absolute return objective, but also identifies the means by which to achieve it. This is exactly why risk parity portfolios are both similar to, and at the same time, different from other alternative investment strategies. Consequently, risk parity strategies are often measured against both absolute and relative return benchmarks. Talk about serving two masters! Another consequence is that because return sources are quite explicit, it is thus possible to assign some numeracy in the benchmarks. Let us go over them one by one.

5.3.3 Sharpe Ratio

We discuss a Sharpe ratio benchmark first. The cash benchmark is a corollary of the expected Sharpe ratio. One of the important results from portfolio theory regarding risk parity is that when all asset classes have the same Sharpe ratio, and their returns have the same correlation, a risk parity portfolio is optimal with the maximum Sharpe ratio. In reality, of course, these conditions are often not met, but historical evidences also suggest they are not terribly wrong and can serve as a starting point for long-term assumptions.

If we include in a risk parity portfolio three primary risk premiums, equity, interest rate, and inflation and assume they are uncorrelated with an expected Sharpe ratio of 0.3, then the Sharpe ratio of a risk parity portfolio consisting of these three risk premiums would be $0.3\sqrt{3} \approx 0.52$. Hence, the benchmark could be a Sharpe ratio of 0.5.

While this ratio might be a reasonable benchmark over the long run, it is not going to be a reliable one for periods when the underlying risk premiums, on which risk parity depends, deviate from their long-run assumptions. Substantial differences occur when all three premiums deliver either much higher or much lower than expected Sharpe ratios, or even negative Sharpe ratios. In the latter case, it is all but impossible to achieve a Sharpe ratio of 0.5. In a long-only portfolio, you cannot make an overall Sharpe ratio positive out of negative Sharpe ratios.

For instance, no risk parity portfolio could have had a positive Sharpe ratio in 2008 because equities and commodities had significantly negative Sharpe ratios. On the other hand, no risk parity portfolio should have a Sharpe ratio lower than one from 2009 to 2012 because all assets' returns delivered positive and, in the case of bonds, significantly positive Sharpe ratios. In summary, the Sharpe ratio can only be used as a long-run benchmark, perhaps with a 5-year window, at the least.

5.3.4 Cash Plus

The excess return of a portfolio is the product of its Sharpe ratio times the portfolio risk. One of the major portfolio characteristics, and indeed a benefit, is that many passive risk parity portfolios target a constant portfolio risk by adjusting leverage ratios. If we target a 10% risk level with a benchmark Sharp ratio of 0.5, then the excess return target would be 5%. Similar to the Sharpe ratio benchmark, the cash plus benchmark can be meaningless in the short run. By definition, the portfolio risk, or "tracking error," between a risk parity portfolio and the cash benchmark would be 10%, which is too large to be considered as intended active risk.

Another issue that is unique to the current zero interest rate environment is that cash plus 5% is simply 5%, which is below the required return level of many institutional investors. This was not an issue when the cash return was 2%–3% in the past. To achieve 7%–8% total return with cash rate of zero, investors need to re-evaluate their Sharpe ratio assumptions, or add other return sources, such as active risk parity, alternative beta in underlying asset class exposures, alpha strategies, and other asset classes with different, unrelated return premiums. Of course, this potential return shortfall is not unique to risk parity portfolios and we shall not discuss them further here.

5.3.5 The 60/40 Benchmark

There are many reasons for this awkward mixing of the old and the new, or more specifically, capital-based benchmarks for risk-based portfolios. First, old habits never die. Almost all institutional investors, on a policy level, still carry capital-based benchmarks. Second, both risk parity and 60/40 are asset allocation portfolios and a risk parity portfolio targeting a constant volatility of 10% will deliver similar return volatility to that of a 60/40 portfolio. As a result, many investors view the allocation to risk parity either as an active decision away from 60/40 portfolios, or as a test of the risk parity concept. Thus, using 60/40 portfolios as benchmarks is a way to keep score for the decision, or to validate the concept.

A typical choice would be the 60/40 portfolio consisting of a 60% capital allocation to the MSCI World Index and a 40% capital allocation to the WGBI Index. Even though it serves as a relative benchmark, it suffers the same short-term problems associated with the Sharpe ratio and cash benchmark. Risk parity portfolios have a large tracking error, typically in the range of 6%–10%, to the 60/40 portfolio. Therefore, annual return differences between a risk parity portfolio and the benchmark are large and, in some years, it could reach 15% or more. These return deviations are mostly due to the fundamental differences between risk parity portfolios and the traditional 60/40 benchmark. In this regard, it could appear that all risk parity strategies perform similarly, either brilliantly or poorly, relative to the benchmark, regardless of their specific or idiosyncratic approaches. In order to assess how different risk parity strategies performed relative to each other, we need a benchmark that captures the basic and essential building blocks of a risk parity-based investment approach.

5.3.6 A Simple Risk Parity Benchmark

We now propose a simple risk parity benchmark with the following ingredients. First, it is based on three primary risk premiums represented by the MSCI World, WGBI, and GSCI Commodity* indices. Second, the risk allocations to the three risk premiums are equal. Finally, the portfolio weights are static and they are determined by the inverse of index return volatilities and total portfolio risk.

The justification of using just these three-asset classes is twofold. One is they represent three primary risk premiums and the other is many other liquid asset classes are, to a large degree, combinations of these three premiums. The equal risk allocation scheme is rooted in simplicity and is not about maintaining a mathematical identity. It is important to note, however, that under certain conditions this equal risk allocation is optimal. We have omitted correlations between the asset classes in determining asset weights; the resulting portfolio is

* Another choice for commodity is DJ-UBS commodity index, which was renamed as Bloomberg Commodity index on July 1, 2014.

TABLE 5.10 Volatility Values and Portfolio Weights for the Three-Asset Classes and also the Portfolio Leverage and Realized Risk

	GSCI (%)	MSCI (%)	WGBI (%)	Portfolio (%)
Volatility	20	15	3.5	8.4
Weight	25	33	142	200

often referred to as naïve risk parity, which is "true" risk parity only when all pair-wise correlations are the same. However, it is empirically true that over the long run the correlations among the three-asset classes will be low. In addition, if we included correlations in the benchmark construction, then we would be forced to choose a specific quantitative method for calculating correlations. For simplicity, leaving it out is probably a better solution.

We cannot avoid being specific entirely. In order to determine the index weights, we must specify index return volatilities. By making them static, we avoid numerical calculations; however, we cannot totally free ourselves from historical hindsight. Qualitatively, the commodity index should have the highest volatility followed by the equity index, leaving the bond index with the lowest volatility. Even though the portfolio risk/return characteristics are not highly sensitive to the exact values, settling on those values is a subjective decision.

Table 5.10 below lists our choices of the volatilities of indices and the resulting weights in the risk parity benchmark. The volatility estimate we choose is 20%, 15%, and 3.5% for GSCI, MSCI, and WGBI, respectively. The weights are inversely proportional to the volatilities and they are scaled to 25%, 33%, and 142% correspondingly so that the resulting leverage is 200%. We shall refer to this portfolio as the risk parity benchmark hereafter.

5.3.7 Characteristics of the Risk Parity Benchmark

We first note the difference between the weights of the risk parity benchmark and the 60/40 benchmark. Relative to the 60/40 benchmark, the risk parity benchmark is long 25% commodities and 102% government bonds while short 27% equities, resulting in a large tracking error. Table 5.11 displays the annualized excess returns and volatilities of the two benchmarks. While the risks are similar, the risk parity benchmark delivered higher returns and a higher Sharpe ratio. In relative terms, the tracking error of the risk parity benchmark is 6.3% and the information ratio is 0.5.

Next, we consider the realized risks and Sharpe ratios of the two benchmarks.* Figure 5.4 plots the 5-year annualized risks. Even though both show variability over time, the risk parity benchmark is less volatile. In other words, the volatility of volatility is lower. On the

TABLE 5.11 Excess Return and Risk of the Two Benchmarks and Their Difference from 1983 to 2012

	60/40	Risk Parity Benchmark	Relative
Excess return	3.1%	6.6%	3.2%
Risk	8.8%	8.4%	6.3%
Sharpe ratio	0.35	0.79	0.50

* Hereafter, performance of both the risk parity and the 60/40 portfolios has been updated to the end of 2014.

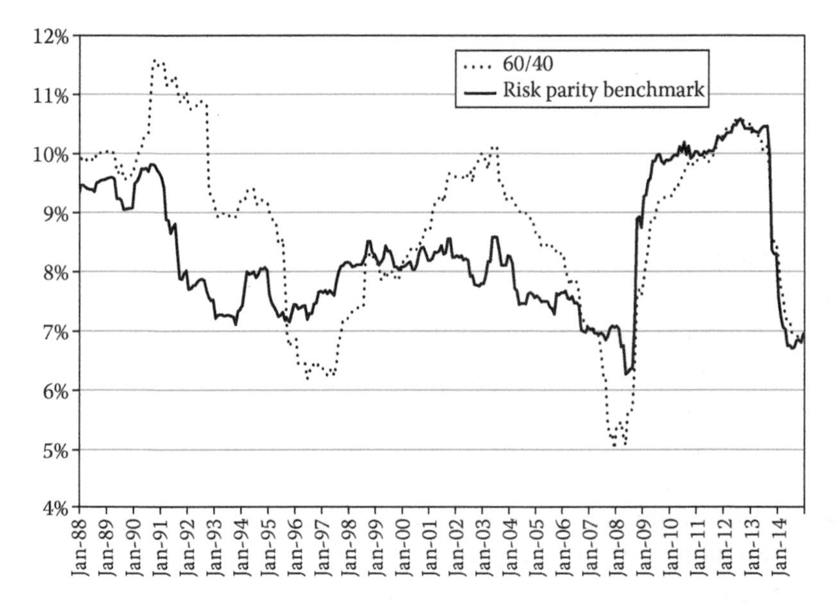

FIGURE 5.4 Annualized risk of rolling 5-year monthly returns.

other hand, the 60/40 benchmark's risk was always very high during the recessionary period but fell to 5% just prior to the financial crisis. Since risk parity portfolios typically target a given risk level, it is preferable to have a benchmark with a tighter risk range.

The same is true for the Sharpe ratios. Figure 5.5 shows the 5-year Sharpe ratios of the two benchmarks. We make two observations. First, not only does the risk parity benchmark have a higher overall Sharpe ratio, it has had less variability over time, ranging from −0.2 to 1.7. The 60/40 benchmark's Sharpe ratio follows closely to that of equities, ranging

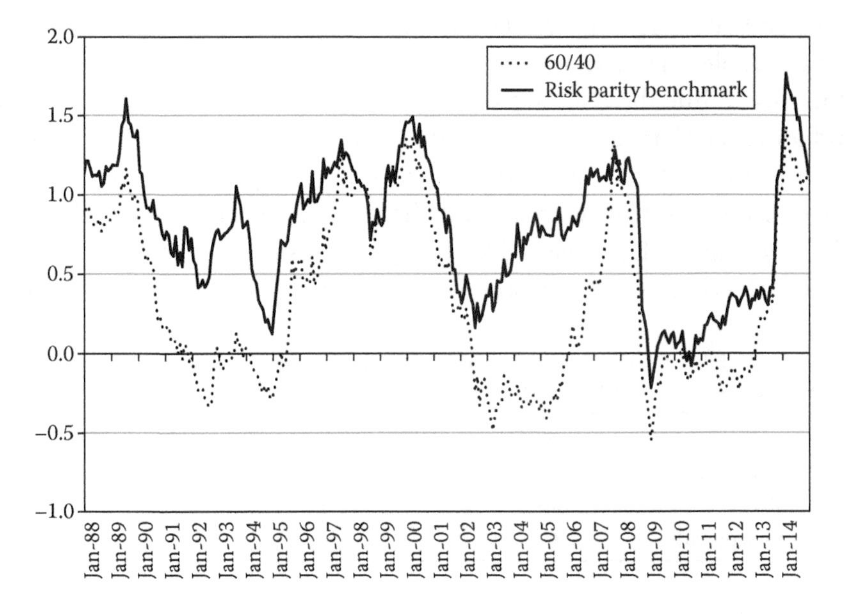

FIGURE 5.5 5-year rolling Sharpe ratio of the two benchmarks.

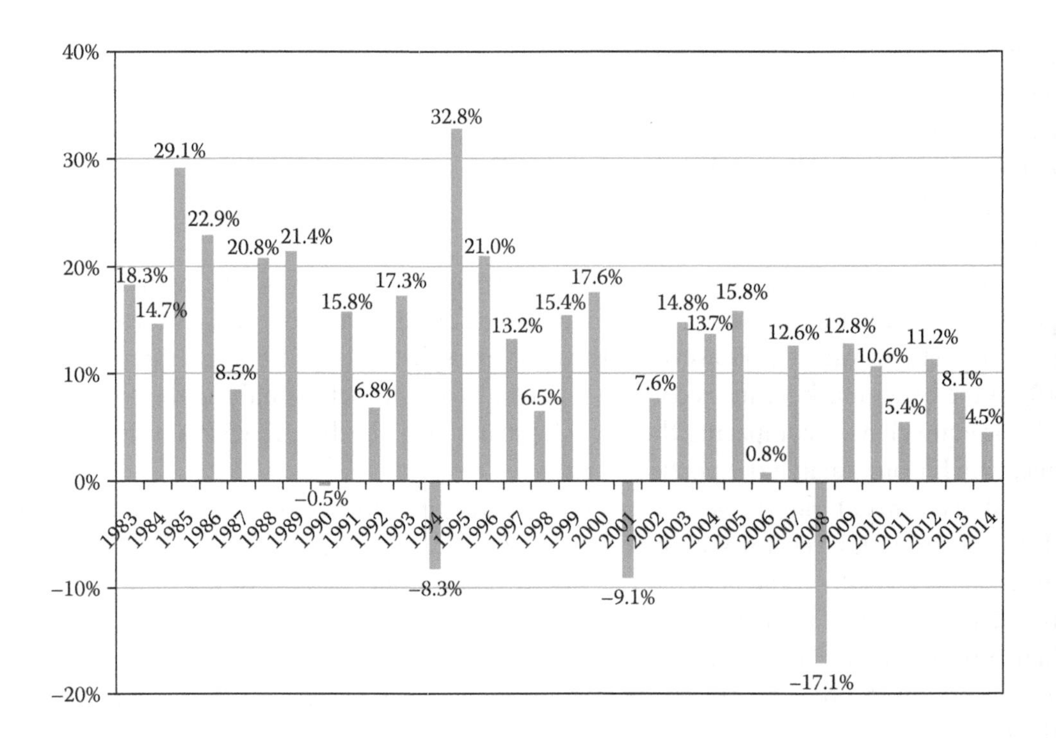

FIGURE 5.6 Annual returns of the risk parity benchmark.

from −0.5 to 1.4. Second, the 5-year rolling Sharpe ratio of the risk parity benchmark shows significant changes over time. One therefore wonders whether a Sharpe ratio of 0.5 over a 5-year horizon is appropriate. Perhaps, an even longer window should be used when considering a Sharpe ratio benchmark.

Finally, we look at the annual total returns of the risk parity benchmark from 1983 to 2014. Over this period, the cash return, indicated by 3-month T-bills, is a little over 4%, so the annualized return of the risk parity benchmark is about 11%. As shown in Figure 5.6, the benchmark returns were mostly positive. The three negative years of 1994, 2001, and 2008 were periods in which either one or two assets had significantly negative risk-adjusted returns. In 1994, it was the bonds; and in both 2001 and 2008, it was equities and commodities with negative risk-adjusted returns. However, in most years the diversification of the three return premiums helped the risk parity benchmark deliver positive returns. In this sense, the benchmark has captured the essence of risk parity investing.

5.3.8 Conclusion

The four possible benchmarks examined, the cash plus, the Sharpe ratio, and the 60/40 benchmarks are only useful and appropriate as long-term benchmarks for risk parity, because the return differences between them and risk parity portfolios are large. But long-term is relative: it could mean 5 years or it could mean 10 years. Based on the evidence presented in this chapter, 5 years may not be long enough.

The proposed risk parity benchmark—based on a naïve construction process that omits correlation, a specified volatility ratio and 200% leverage—can serve as a short-term relative benchmark for risk parity portfolios. It is also true that short-term can mean different things to different people. First, we can measure risk parity portfolios against the benchmark in terms of Sharpe ratio. This has the benefit of equalizing differences in the risk levels of portfolios and their benchmark. This comparison can be done at least on an annual basis since one needs a string of returns to calculate Sharpe ratios. Therefore, here short-term is 1 year.

What about the risk parity benchmark as a return benchmark? If the realized risk of a portfolio is similar to that of the benchmark over a specific period, say a year, we can compare the returns directly. However, if the risks are substantially different, then we should adjust the benchmark return according to the ratio of risks before making the comparison. This is equivalent to comparing the portfolio to a scaled benchmark with a different leverage, which would match the portfolio's risk.

If short term is 1 month, we can only measure relative returns. If one does not care about difference in risk levels for a month, we can simply use the raw returns. However, if one does care, especially if the targeted risk of a given portfolio is much different from 8.4%—the historical risk of the benchmark—one can make a similar risk-based adjustment to the benchmark return. These adjustments, either on a monthly basis or on an annual basis, are necessary in order to derive the risk-adjusted alphas of risk parity portfolios versus the risk parity benchmark. It can be recognized that the same approach has been used in portfolio attribution for many active strategies. In active equity, the adjustment is the beta of a portfolio versus its capital-weighted index. In active fixed income, the corresponding effect is the duration.

A proper benchmark is crucial for analyzing and understanding any investment strategy. This chapter presents a simple but intuitive benchmark that captures the essence of risk parity investing. It is the first, but an important step in that direction. It serves as a benchmark and in addition, it can be used in performance attribution, together with other factors that are relevant to the relative performance of specific risk parity strategies.

5.4 UPSIDE PARTICIPATION AND DOWNSIDE PROTECTION AND RISK PARITY PORTFOLIOS*

5.4.1 The New Holy Grail

"Upside participation and downside protection" has always been one of the qualitative or otherwise loosely defined goals of active investing. With the global financial crisis of 2008 "permanently" etched on the psyche of the investment community, this goal, especially the latter half of it, has taken on much more significance in recent years. To many investors and managers, this has become the new Holy Grail of investing.

In the pursuit of this Holy Grail, two important questions need to be addressed. The first is how we quantitatively define "upside participation and downside protection." The second question is how we achieve it, and more importantly, how we achieve it without sacrificing long-term returns.

* Originally written by the author in June 2014, with the help of Nicholas Alonso, Mark Barnes, Bryan Belton, Kun Yang.

In a recent article (Qian 2015), I offered some answers to the first question by providing an analytic framework that defines the upside and downside participation ratios of an active strategy relative to a market benchmark. We also use the participation ratio difference (PRD) as a statistics measure for the efficacy of "upside participation and downside protection." We show that a positive PRD is an indicator of "upside participation and downside protection" for a positive PRD implies that the degree with which the active strategy goes down in a down market is less than the degree with which it goes up in an up market. We also proved that the PRD measure is directly related to the alpha of the strategy versus the benchmark with respect to a one-factor CAPM model. In other words, a strategy with a positive PRD is also one with positive alpha and vice versa.

5.4.2 Three Ways to the Holy Grail

We shall provide a brief description of participations and PRD shortly. However, the focus of this research note is about the second question of how to achieve "upside participation and downside protection." In our view, there are at least three approaches to it.

One is to invest in low volatility and/or low beta strategies, such as minimum variance or defensive equity strategies. These strategies drastically reduce portfolio risk and beta so downside participation ratios are quite low. On the other hand, they have historically relied heavily on the low volatility anomaly to deliver enough alpha to outperform capitalization-weighted benchmarks. The potential risk to this approach is if the low volatility anomaly fails to deliver in the future, these strategies will have very low upside participation ratios as well and consequently, low alpha and negative excess returns relative to their benchmarks.

The second approach is to design strategies with tail-risk hedging based on portfolio insurance schemes. The hedging might be provided by either derivatives such as options and swaps or stop-loss policies using trend-based signals traditionally utilized by CTA managers. These strategies became quite popular after the 2008 financial crisis, because with hindsight they worked very well during the crisis for the obvious reason that the type of portfolio insurance offered by these strategies was exactly what the doctor would have ordered. However, many stop-loss investment policies have actually detracted value over the long run due to explicit or implicit costs associated with these strategies (see Chapter 7, Section 7.3) and they have since stopped working in the last several years. In addition, whether they would provide needed insurance in future financial crises remains an open question.

The third category of investment strategies to deliver "upside participation and downside protection" is through the true diversification offered by portfolios like risk parity. Compared to the previous two approaches, the investment objective of risk parity is not explicitly about downside protection. Rather it is intended to capture asset returns or risk premiums in the most diversified way. However, in the process of doing so, as we shall see in this note, it proves to be a very efficient way to deliver "upside participation and downside protection."

Why is this true? To answer this question, one has to take a deep look at traditional market indices. Almost none of the traditional market indices are truly diversified, due to various embedded risk concentrations. For example, equity indices have risk concentration in countries, sectors, and stocks. Fixed-income indices are typically concentrated across the term structure (maturities) and credit ratings. Finally, commodity indices have risk concentrations in sectors and individual commodities. These risk concentrations cause high volatility in market indices, characterized by sharp upturns and deep drawdowns. By diversifying away these risk concentrations, which are in total contradiction with the "passive" labeling of market-weighted indices, risk parity portfolios seek to avoid sharp moves in both up and down markets resulting in positive PRDs. In addition, since volatilities and betas of risk parity portfolios are only slightly lower than that of market indices, risk parity portfolios typically outperform market indices in the long run.

5.4.3 Participation Ratios and PRD

We define participation ratio as the ratio of the conditional mean or average of a strategy versus the conditional mean or average of the corresponding index, conditioned on the sign of the index return relative to cash.

Suppose the index excess return is denoted by r_x and the strategy excess return* is denoted by r_y. Then the upside participation ratio is

$$P_+ = \frac{E(r_y \mid r_x > 0)}{E(r_x \mid r_x > 0)}. \tag{5.4}$$

The notation $E()$ in the equation denotes expectation or average. Similarly, we define the downside participation as

$$P_- = \frac{E(r_y \mid r_x < 0)}{E(r_x \mid r_x < 0)}. \tag{5.5}$$

With both participation ratios defined, we denote the PRD as

$$\text{PRD} = P_+ - P_-. \tag{5.6}$$

When the upside participation ratio is greater than the downside participation ratio, PRD is positive. Naturally, all else equal, one would prefer a strategy with a positive PRD and a positive alpha to one with a negative PRD and a negative alpha. By definition, only strategies with positive PRDs qualify as strategies that provide "upside participation and downside protection."

5.4.4 Participation Ratio and PRD of Risk Parity Portfolios

We now study the participations ratios and PRDs of risk parity portfolios in two parts. The first part will be focused on risk parity portfolios at the asset class level. In the next section, we will move on to risk parity multi-asset portfolios.

---

* The use of excess return instead of actual return is to makes both participation ratios of cash zero as they should be since it really does not participate in any way.

TABLE 5.12 Asset Classes with Corresponding Benchmark Indices

Asset Class	Index
Commodity	GSCI
UST	Barclays UST
US Credit	Barclays US Credit
World ILB	Barclays ILB
World non-US Bond	WGBI ex US
US Large Cap	S&P 500
US Small Cap	R2000
World non-US Equity	MSCI ex US
EM Equity	MSCI EM

Table 5.12 provides the list of nine asset classes and their corresponding market-weighted indices. For each asset class, we shall construct a risk parity portfolio based on our risk parity methodology with either futures or physical securities. In markets where we use futures, we target the volatility of a risk parity portfolio to be consistent with the volatility of the indices. With portfolios of physical securities, we have portfolios that are fully invested without leverage. The backtest returns of all risk parity portfolios span from January 1995 to December 2013 on a monthly basis. Based on the monthly returns of the risk parity portfolios and their corresponding indices,* we calculated upside and downside participation ratios conditioned on the monthly returns of individual indices and the resulting PRDs.†

Table 5.13 shows the participation ratios and PRDs for the nine asset classes, together with beta-adjusted alphas and excess returns versus the indices. The upside participation

TABLE 5.13 Participation Ratios, PRDs, Alphas, and Excess Returns of Risk Parity Asset Class Portfolios

Asset Class	P_+	P_-	PRD	Alpha (%)	Excess Return (%)
Commodity	0.88	0.54	0.34	0.73	0.61
UST	1.02	0.94	0.08	0.05	0.04
US Credit	0.85	0.84	0.01	0.01	−0.04
World ILB	0.95	0.86	0.09	0.04	0.02
World × US Bond	0.89	1.01	−0.12	−0.04	−0.05
US Large Cap	0.94	0.80	0.15	0.25	0.17
US Small Cap	0.90	0.84	0.06	0.16	0.06
World non-US Equity	0.99	0.62	0.36	0.60	0.57
EM Equity	0.87	0.67	0.20	0.52	0.41
Average	0.92	0.79	0.13	0.26	0.20

* We use 3-month UST bill return as cash return.

† The detailed calculation for each pair of risk parity portfolio and its corresponding market index, based on monthly excess returns is as follows. First, we identify all the months during which the index returns were positive. Then we calculate the average returns of both the index and the risk parity portfolio in those months and take the ratio of two averages to obtain the upside participation ratio. Second, we repeat the process for the downside participation ratio by using average returns during the months when the index return is negative.

ratios range from 0.85 to 1.02 with an average of 0.92. The downside participation ratios range from 0.54 to 1.01 with an average of 0.79. Finally, the PRDs are all positive except for that of the World ex US Government Bond Index. They range from −0.12 to 0.36 with an average of 0.13.

We make several remarks about these results. First, the upside participation ratios are, in general, below and close to one. This is consistent with the fact that risk parity portfolios are not necessarily low volatility strategies so they can have strong upside participation. Second, the downside protection is quite strong except for the World ex US bond portfolio. This exception was the result of outstanding performance of the WGBI ex US index whose largest allocation was to JGBs. JGBs were one of the best-performing sovereign bonds during this period due to the disinflation and often outright deflation in Japan. While risk concentration embedded in traditional indices generally does not pay, it can be rewarded on rare occasions.

Some aspects of alpha and excess return are also worth noting. First, there is strong correlation between PRD and alpha, as we have shown theoretically in the aforementioned article. For example, the monthly alpha is positive 73 bps for the risk parity commodity portfolio and negative 4 bps for the World ex US bond portfolio. The average monthly alpha is 26 bps, or close to 3% on an annual basis.

The excess returns, on the other hand, are slightly lower than alpha, because the risk parity portfolios' betas to their indices are less than one and excess returns assume they have a beta of one. As a result, the commodity portfolio's excess return is 61 bps and the World ex US bond's excess return is −5 bps instead of −4 bps. We also note that the excess return of the risk parity credit portfolio is now −4 bps. Compared to the index, the risk parity credit portfolio is higher in credit rating but lower in duration. Given its flat PRD, the relative short duration position largely explains the underperformance. However, excess returns are positive for the other seven asset classes and the average is 20 bps per month. Overall, we can also conclude that in seven out of nine cases, the risk parity portfolio succeeded in achieving the goal of "upside participation and downside protection."

5.4.5 Participation Ratios of Risk Parity Multi-Asset Portfolios

Risk parity multi-asset portfolios utilizing underlying risk parity asset class portfolios achieve "upside participation and downside protection" from both a top-down and bottom-up perspective. On the other hand, risk parity multi-asset portfolios utilizing traditional market indices can only rely on the top-down approach. In this section, we derive the participation ratios of risk parity multi-asset portfolios based on back test results from January 1995 to December 2013 and compare the two implementations. In addition, we shall also construct a third risk parity multi-asset portfolio, which incorporates a dynamic risk allocation approach and demonstrate that it further improves its ability to provide "upside participation and downside protection."

We must first choose a benchmark for the multi-asset portfolios. Traditional asset allocation based on notional allocation rather than risk allocation must have significant equity and equity-like assets to generate sufficient long-term returns. Many institutional investors have created policy portfolios that resemble a 60/40 portfolio with 60% in equity and 40%

TABLE 5.14 Annualized Average Excess Return, Standard Deviation, and Sharpe Ratio of Multi-Asset Portfolios

	Average Return (%)	Standard Deviation (%)	Sharpe Ratio
60/40	4.15	8.99	0.46
RPMA I	8.02	9.24	0.87
RPMA II	9.46	8.94	1.06
RPMA II DRA	11.49	8.67	1.33

in fixed income. Even though institutional investors have diversified away from 60/40 into other alternative investments such as private equity, real estate, and hedge funds, it can be argued that their portfolios still mirrors that of 60/40 from the perspective of aggregated risk allocation. For instance, their portfolio returns during the 2008 financial crisis are a case in point.

We thus construct a 60/40 portfolio based on the equity and fixed-income asset classes listed in Table 5.12. The 60% equity allocation consists of a 25% allocation to both the S&P 500 index and MSCI ex US index, as well as a 5% allocation to both the Russell 2000 index and the MSCI EM index. The 40% fixed-income allocation consists of a 15% allocation to both the UST index and the WGBI ex US index, and a 5% allocation to both the credit index and the inflation-linked bond index.

Table 5.14 displays the annualized return statistics of the 60/40 benchmark as well as the three risk parity multi-asset portfolios. The annualized return volatilities are all around 9% while the returns exhibit significant differences. The biggest return improvement is from 60/40 to RPMA I, which is a risk parity portfolio constructed using traditional market indices. The return almost doubled from 4.15% to 8.02%. The other two incremental improvements are also meaningful. RPMA II, which applies the risk parity principle both across as well as within asset classes, improves the return by almost 150 bps per year. Finally, the portfolio that incorporates dynamic risk allocation by applying tactical shifts to the portfolio's strategic risk allocation (RPMA II DRA) enhances return by an additional 200 bps over RPMA II. How did risk parity multi-asset deliver such strong performance relative to the 60/40 portfolios? The participation ratios help to answer this question.

Table 5.15 shows the participation ratios and PRDs of the three risk parity multi-asset portfolios, together with beta-adjusted alphas and excess returns. The upside participation ratios are all close to one while the downside participation ratios are much less than one. Hence, the answer to the performance question is that risk parity portfolios had strong downside protection in bad times while keeping up with the 60/40 portfolio in good times.

TABLE 5.15 Participation Ratios, PRDs, Alphas, and Excess Returns of Risk Parity Multi-Asset Portfolios

	P_+	P_-	PRD	Alpha (%)	Excess Return (%)
60/40	1.00	1.00	0.00		
RPMA I	0.94	0.53	0.41	0.42	0.32
RPMA II	0.93	0.37	0.56	0.57	0.44
RPMA II DRA	0.98	0.24	0.73	0.76	0.61

Intuitively, it is easy to understand why this is the case. When the 60/40 portfolio has a positive return, it is mostly due to a positive equity return since the 60/40 portfolio is dominated by equity risk. Risk parity portfolios have significant exposure to equity risk, so they will participate strongly in an equity rally. However, despite having significant equity exposure, it is much less than that of an equity centric 60/40 portfolio. As a result, periods of concentrated equity rallies are likely to leave large return gaps between the risk parity portfolios and the 60/40 portfolio. During these periods, exposure to commodities within risk parity portfolios helps to minimize the return gap and thus raise the upside participation ratios close to one.

When the 60/40 portfolio has a negative return, it is mostly due to a negative equity market performance. In this case, the significant exposure to high-quality fixed income provides true diversification that offsets the loss contribution from equities. This offset either reduces the portfolio losses or turns a profit for risk parity portfolios. The aggregated effect is a much smaller loss for the risk parity portfolios, that is, strong downside protection.

Another important feature of Table 5.15 is that the downside participation ratios get progressively smaller as we move from RPMA I, to RPMA II, to RPMA II DRA. The improvement of RPMA II over RPMA I is due to the increased downside protection attributable to the underlying risk parity asset class exposures. For example, all four risk parity equity portfolios have downside participation ratios less than one (see Table 5.13). In the case of World ex US and EM equity, the ratios are significantly less than one. When equity markets are down, causing losses for the 60/40 portfolio, these risk parity equity portfolios provide additional downside protection for RPMA II and RPMA II DRA.

The effect of dynamic risk allocation is twofold. Not only does it improve downside protection, but it also improves upside participation. Comparing RPMA II DRA to RPMA II, we see the upside participation ratio goes up from 0.93 to 0.98 and the downside participation ratio goes down from 0.37 to 0.24. This is also intuitive since a valuable tactical process is supposed to add value in both up and down markets.

The participation ratios result in significant positive PRD for the risk parity portfolios and significant alphas and excess returns versus the 60/40 portfolio. In Table 5.15, the alphas and excess returns are expressed on a monthly basis for consistency with participation ratios. On an annual basis, the excess returns are 3.9%, 5.3%, and 7.3%, respectively, for RPMA I, RPMA II, and RPMA II DRA portfolios.

5.4.6 Conclusion

"Upside participation and downside protection" is neither an elusive goal nor a mere marketing slogan. It can be rigorously defined in terms of quantitative return and risk measures. In this chapter, we have shown that one concrete approach to achieve this goal is through true diversification. Risk parity, both across and within asset classes provides significant downside protection without forgoing much upside participation relative to a 60/40 benchmark. The combination of both risk parity applications results in a major PRD and a higher excess return over the 60/40 portfolio. We also point out that a dynamic risk allocation process can further improve both upside participation and downside protection.

Risk parity accomplishes "upside participation and downside protection" through a unified theme of true diversification. Unlike other approaches such as minimum variance and stop-loss policies, the success of risk parity does not rely on any particular factor or forecasting signal, whose efficacy can disappear with changing market conditions and changing investor behavior. Rather, the real strength of risk parity lies in its simplicity, true passivity, and indifference to the crowds, that is, the traditional market indices.

yields. However, as I wrote previously, placing too much emphasis on individual asset classes in the context of a diversified portfolio misses the point (see Chapter 4). Indeed, with risk parity, the risk of rising yields is the same as the risk of falling equity markets; that is why the two risks are balanced. By the same token, one should not focus on a single year to decide the merits of an investment strategy.

On the other hand, however, the memory of 1994, either true or distorted, does raise the concern about rising bond yields to a new level. With growing confidence in the US economy and continued monetary easing across the globe, equities have rallied and sovereign bond yields have risen since the second half of 2012. If the trend indeed continues, bond yields might rise further.

However, before getting carried away with fear, let us review what really happened in 1994 and decide whether the current market environment actually resembles that of 1994. In this investment insight, I show that while risk parity had negative returns in 1994 the loss was rather modest. I then argue that the current macroeconomic conditions are very different from the conditions that existed in 1994, which makes significantly higher yields in 2013 a remote possibility.

6.1.1 What Happened in 1994?

6.1.1.1 Interest Rates

There is no question that, the culprit of the bond market "crash" in 1994 was the Fed. The starting point for discussing the events of 1994 is interest rates.

Figure 6.1 shows the Fed funds target rate from 1990 to 1994. Following the 1990/1991 recession, the Fed cut the short-term rate by more than 500 basis points and kept the rate at 3% throughout 1993. Starting in early 1994, and continuing through the year, the Fed unexpectedly raised the target rate six times with a cumulative increase of 250 basis points.

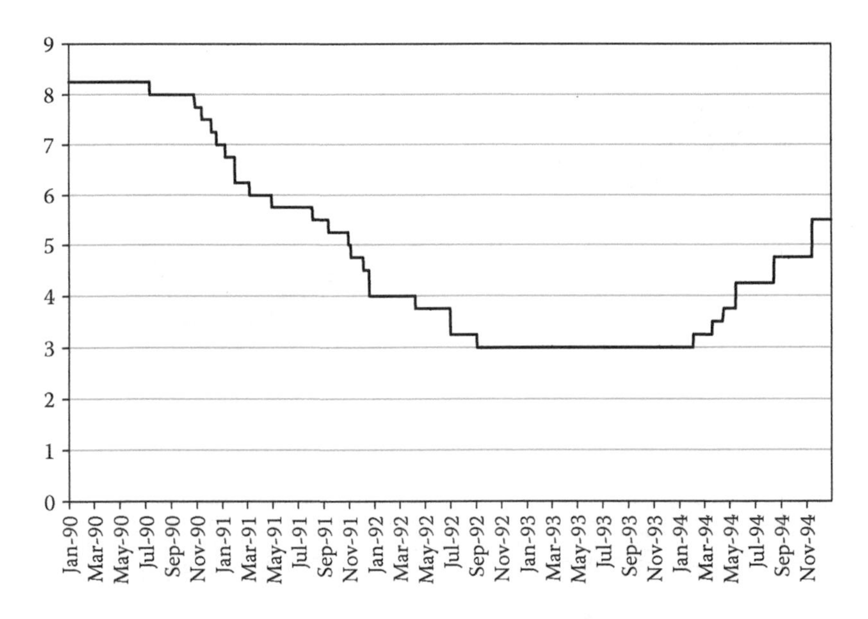

FIGURE 6.1 The Fed fund target rate from 1990 to 1994.

FIGURE 6.2 US 3-month T-bill rate from 1990 to 1994.

What happened to the short-term interest rate is rather predictable—it followed the Fed funds rate closely. Shown in Figure 6.2, the 3-month rate went from around 3% to 5.7%. So, the "good" news was the cash return rose in 1994.

The bad news was long-term yields also rose significantly. However, the long-end of the yield curve, also influenced by inflation expectations and expected term premium, was slightly out of sync with short-term rates. Figure 6.3 shows the 10-year treasury yield, which jumped by almost 200 basis points, from about 5.5% to 7.5%, during the first months of the rate hikes. Then its ascent paused, and it ended the year at 7.8%.

FIGURE 6.3 US 10-year treasury bond yield from 1990 to 1994.

FIGURE 6.4 Yield curve slope (10 y/3 m) from 1990 to 1994.

The differing dynamics of short- and long-term yields, in response to the Fed funds rate, are reflected in the slope of the yield curve. Figure 6.4 shows the yield spread between the 10-year yield and the 3-month rate. The treasury market initially went through a bear steepening with the spread going from 250 to 350 bps. However, the yield curve subsequently flattened with the spread falling to around 200 bps.* Near the end of 1994, the bond market—as reflected by the term structure of interest rates—had started to price in lower inflation and lower growth expectations.

6.1.1.2 Bond Returns

The reason we highlight the yield curve dynamics above is that the change in the shape of the yield curve affects the bond returns along the curve. With a flattening yield curve, the impact of rising yields will be less for bonds with longer maturities on a duration-adjusted basis.

Table 6.1 displays the Citigroup treasury index return and subindex returns with different maturity buckets in 1994. The cash return from 3-month T-bills was close to 4% and all other index returns were below the cash return. The aggregate index return was −3.4%, while the subindex returns ranged from 0.5% for short-term bonds to −8.2% for long-term bonds.

TABLE 6.1 Citigroup UST Index Returns in 1994

	Cash (%)	**1–3 years (%)**	**3–7 years (%)**	**7–10 years (%)**	**20 years+ (%)**	**Index (%)**
Return	4.2	0.5	−3.1	−5.7	−8.2	−3.4

* The long-term average yield spread is about 200 bps and the current level (at the end of March 2013) is close to the average.

TABLE 6.2 Excess Return, Volatility, and Sharpe Ratio of Citigroup UST Indices

	1–3 years	3–7 years	7–10 years	20 years+	Index
Excess return	−3.6%	−7.0%	−9.6%	−12.0%	−7.3%
Risk	1.6%	3.6%	5.8%	8.6%	4.2%
Sharpe	−2.28	−1.96	−1.66	−1.38	−1.75

On a risk-adjusted basis, however, the long-term bond bucket (20 years+) was actually the best among the group due to the flattening of the yield curve. Table 6.2 shows excess returns over cash, risk (annualized monthly return standard deviation in 1994), and Sharpe ratios of all indices. The index for the longest maturity bucket (20 years+) had a Sharpe ratio of −1.38 while short-term bonds fared much worse, with a Sharpe ratio of −2.28.*

In such an environment, investors are not entirely helpless. For example, shifting a portfolio's risk allocation from short- to long-term bonds and at the same time reducing the overall exposure would mitigate some negative performance. While extending the average maturity of a portfolio during a period of rising interest rates may appear to be counterintuitive, it is actually quite sensible in bear flattening environments if it is done without changing the aggregate risk allocated to fixed income. Indeed, we find that applying a dynamic risk allocation methodology would have helped reduce the negative contribution from fixed income in 1994.

6.1.1.3 Returns of Other Asset Classes

During 1994, few asset classes were spared from the rising interest rates. Equities did not perform well: as displayed in Table 6.3, the MSCI World Index was down by 2.2%. In addition, equity returns were down for both small cap as well as emerging market stocks. Returns were also negative for many other fixed-income assets. The WGBI was down 3.7%. The one bright spot in 1994 was the commodity market. The DJ-UBS index earned a positive return of 16.6%. In hindsight, the Fed's rate hikes were preemptive moves aimed to slow down economic growth and inflation, which was partly reflected in rising commodities prices.

6.1.1.4 Returns of Risk Parity Portfolios

With negative returns for both equities and bonds in 1994, any asset allocation portfolio would have been hard pressed to generate positive returns. As shown in Table 6.4, the traditional 60/40 portfolio with 60% of its capital invested in the MSCI World Index and 40%

TABLE 6.3 Returns of Commodity, Equity, and Bond Indices in 1994

	DJ-UBS (%)	MSCI World (%)	WGBI (%)
Return	16.6	−2.2	−3.7

* It is certainly debatable whether these negative returns are qualified as crashes. On an absolute basis, the index return of −3.4% (Table 6.1) represents a minor loss. However, an excess return of −7.3% is a significant loss relative to cash. If equities had a Sharpe ratio of −1.75, assuming a risk of 15%, then the implied excess return would have been −26.3%, which is firmly in bear market territory for equities. Perhaps, a bear market is the right description for the bond market of 1994 since the loss was accumulated over a period of several months.

TABLE 6.4 Absolute Returns of a 60/40 Portfolio and a Static Risk Parity Portfolio

	60/40 (%)	Risk Parity (%)
Return 1994	−2.7	−5.9
Return 2008	−22.9	−14.0

of its capital invested in the WGBI returned −2.7% in 1994 while a static risk parity portfolio returned −5.9% in 1994. The static weights are based on a naïve risk parity weighting* scheme with equal risk contribution from three sources of returns: equities, interest rates, and commodities.

Given these explicit historical returns, we can now make several comments. First, while the returns in 1994 are negative, the portfolio losses are by no means disastrous. For comparison purposes, Table 6.4 also shows the returns of two portfolios in 2008. The losses of 2008 were much larger, especially devastating to the 60/40 portfolio.

Second, from a risk-adjusted perspective the portfolio's performance also seems ordinary. Both portfolios tend to have an annual volatility around 10%. From that perspective the loss experienced in 1994 by the risk parity portfolio does not even amount to a one-standard deviation event. This is, in no small measure, due to the positive contribution from commodities, which offsets some of the negative contribution from bonds. In the sense of diversification, risk parity worked well in limiting potential losses in 1994.

Finally, we note that one can improve the static risk parity portfolio through both further diversification and dynamic risk allocation. For example, in 1994, dynamic risk allocation would have led to a gradual increase in the risk allocated to commodities accompanied by a gradual decrease in the risk allocated to nominal fixed-income assets. Our research has found that these enhancements can offer significant downside protection to a static risk parity portfolio.

6.1.2 Is 2013 Like 1994?

I do not confess to know whether the rest of 2013 will turn out to be like 1994 in terms of asset returns. So far, it has not. Equities have rallied, treasury yields have been range bound, and commodity prices have been stagnant. However, we do know that currently the macroeconomic environment shares few commonalities with that of 1994.

First, it is a near certainty that the Fed shall not raise the Fed Funds rate as they did in 1994. As evidenced by Bernanke's strong bias to signal his policy stance, the Fed has learned that it is unwise to shock the markets. In addition, the forward rate guidance provided by the Fed strongly indicates that the ZIRP will be kept until the US unemployment rate dips below 6.5%. As of March 2013, it is at 7.7% and the Fed does not expect the rate to fall below 6.5% until sometime in 2015.

One might argue that monetary tightening could come if the Fed stops the QE program. This is also unlikely in 2013 based on the latest economic data and the Fed policy statements. Over the last several years, contrary to common belief, treasury yields have fallen

* See Section 5.3 in Chapter 5. We have substituted DJ-UBS index for the GSCI index, which was used for its longer return history. The weights are 25% in DJ-UBS index, 33% in MSCI index, and 142% in WGBI index, respectively.

every time the Fed finished its preannounced calendar-based asset purchasing programs. This is because the end of QE or even not re-investing the interest proceeds from the assets already purchased is viewed by the market as monetary tightening, which caused long-term rates to fall.

Another important but subtle difference between now and 1994 is the transparency of the Fed policy. The rate hikes of 1994 were not expected by many market participants at the time, and as a result, the market reactions were quite volatile. In contrast, the Fed policy has become more transparent under Chairman Bernanke. It is reasonable to expect that the path of monetary policy will be more gradual and perhaps more predicated on economic data.

Assuming this time the Fed continues the "open-ended" QE for the remainder of the year, the long-end of the yield curve could move higher if growth prospects and/or inflation expectations move higher. This brings us to the third major difference between now and 1994.

Figure 6.5 shows US GDP growth from 1990 to 1994. The growth rate exceeded 4% for three consecutive quarters from the end of 1993 to the first half of 1994. This is one of the major reasons for the rate hikes in 1994. In contrast, the current growth in the United States hovers around 2%, with a high degree of uncertainty caused by persistent deleveraging pressure and recurring instability out of Europe.

The inflation story is similar but somewhat less dramatic. Figure 6.6 shows the year-over-year change in core CPI. It had been trending down from 1991 to 1993. Perhaps, this is why the bond market was caught off guard by the Fed's actions in 1994. Still, the inflation rate was well above 2% and there was a slight pick-up at the beginning of 1994. Currently, the inflation rate is near 2% and the inflation breakeven rate is near 2.5%. As long as inflation expectations are stable, the Fed likely will not change its easing policy.

FIGURE 6.5 Q2Q annualized GDP growth from 1990 to 1994.

FIGURE 6.6 Year over year change of the US CPI index.

Finally, the global economy and capital markets continue to work through the aftermath of the global financial crisis of 2008. The situation is markedly different from that of 1994, when the US economy had recovered sufficiently from a traditional demand-driven recession of 1990/1991. If anything, the year of 2013 will probably be more like the year of 2012 than the year of 1994.

6.1.3 Conclusion

In finance or economics, historical facts cannot change. However, the same events or information often evoke different, and sometimes completely opposite reactions and interpretations from different people—the root of cognitive bias in behavior theory. Could the year of 1994 be such an event, the interpretation of which highly depends on views one already held regarding interest rates?

If one is to think that bond yields are about to shoot up significantly, the year of 1994 seems to support the hypothesis. Furthermore, the potential loss to risk parity can dominate thinking. On the other hand, if one believes that long-term bond yields are mostly determined by economic conditions and monetary policy, then the year of 1994 seems rather irrelevant to the current situation.

In this research note, I have argued for the second view. First, I show that potential losses incurred by risk parity portfolios in 1994 are quite manageable. The risk of rising yields to a risk parity portfolio is not as horrific as some fear. In addition, we need to be mindful that fear can lead to faulty perceptions and poor judgment. Second, I argue that the current macroeconomic conditions are dissimilar to those that existed in 1994.

Nevertheless, if one were to draw the conclusion that bond yields won't rise at all in 2013, it could well be another form of cognitive bias. Certainly, bond yields could rise if growth prospects improve and the Fed continues to maintain the "open-ended" QE program.* If that were the case, then the increase in bond yields would be limited. Risky assets such as equities and commodities would perform well enough to offset losses by fixed-income assets, giving risk parity a modestly positive return. On the other hand, bond yields could fall, if global growth, especially the growth in the United States, falters again. Then equities and commodities would suffer losses. These possible outcomes lead us back to risk parity, which balances these different types of risks to form a truly diversified portfolio. If bonds yields were to rise in the future like in 1994, or in any other periods of rising yields, dynamic risk allocation allows us to shift the risk allocation to provide further downside protection against such risks.

6.1.4 Postmortem

It turned out that in 2013 capital markets did experience a volatile period in May and June, known as the "taper tantrum," when the former Fed Chairman Ben Bernarke hinted that QE would taper off in not-so-distant future (Bernanke, 2013). Global bond yields, especially those in the United States rose significantly, and both equities and commodities sold off in those 2 months. As a result, risk parity portfolios experienced a sharp drawdown in May and June 2013. However, capital markets soon recovered with strong equity returns and stable bond prices. We have a more detailed discussion on this event in the next section. For the whole year, the returns were 26.1%, 0.2%, and −9.5% for the MSCI World Index, WGBI index, and DJ-UBS index, respectively. The risk parity portfolio based on these three indices had a return of 5.75%. Rather than the year of the bond market crash, 2013 might be characterized as the year of an equity market boom. Well-constructed risk parity portfolios would benefit from its exposure to equity risk premium and deliver positive returns.

6.2 AFTER TAPER TANTRUM: AN IMPROVED OUTLOOK FOR RISK PARITY†

Risk parity multi-asset portfolios suffered large losses during the months of May and June 2013 when the Fed hinted at a tapering of its QE program. Fixed-income assets led the way down with rising yields. Equities put up a good fight in May but eventually succumbed in June. Commodities fared slightly better, only because the energy sector rose under the cloud of the Syrian conflict. Risk parity multi-asset portfolios invest in these three broad asset classes with a balanced risk allocation to achieve diversification. When all asset classes had rather sizeable negative returns in those 2 months, diversification was absent, resulting in large losses for risk parity portfolios.

Since June, equities and commodities have recovered most of the losses and bond yields seem to have stabilized. As a result, the performance of risk parity portfolios has improved. But what does the future hold for risk parity? While it is hard to predict the future—this is partly the reason for risk parity investing, there are a few telltale signs that the future is likely to be brighter.

* Contrary to popular belief, the impact of QE on long-term bond yields is not one directional. While there is downward pressure on yields due to the Fed's asset purchasing program, the stimulus effect of QE also cause yields to rise.

† Originally written by the author in September 2013.

Here I present three arguments for this thesis. First, there are historical precedents where risk parity portfolios recovered quite strongly after episodes of assets repricing, caused by either Fed policy changes or market panics. Second, the steepness of the yield curve suggests that fixed-income assets are more reasonably priced today than they were several months ago and it leads one to expect positive excess returns. Third, correlations among various assets classes have dropped from very high levels reached during the second quarter of this year. It is likely they will stay at the current levels or even turn negative again, and hence restore risk parity's diversification benefit.

6.2.1 Historical Precedents

Risk parity portfolios provide true diversification by balancing risk allocations to different risk premiums and/or asset classes. While this is theoretically indisputable and empirically verified over long time horizons, diversification benefits can be temporarily absent in the short term. Both changes in monetary policy and broad risk aversion can cause simultaneous declines of many asset classes. When this is the case, it is as though investors all rush to cash, possibly because cash might become more valuable (higher risk-free rates in the future), or cash becomes the only safe asset to investors at that moment. In both cases, however, it can be argued that the risk premiums of many assets have risen, which bodes well for the future expected returns of these assets. In other words, large losses caused by rising risk premiums, rather than macroeconomic fundamentals such as growth and inflation sow the seed of future recovery.

Since 1990, there are three instances where many asset classes declined because the Fed was expected to tighten monetary policy: one in 1994, one in 2004, and one in 2013. In 1994, the Fed surprised the market by raising the Fed funds rate seven times with a cumulative increase of 300 basis points. As a result, both equity and fixed-income assets performed poorly while commodities delivered positive returns. Most of the portfolio losses experienced in 1994 occurred between February and April. Our backtesting results shown in Table 6.5 suggest our risk parity portfolio declined –10.1% during those 3 months. However, the portfolio rebounded nicely delivering a return of 15.1% for the next 12 months as the initial shocks wore off and both equity and fixed-income assets recovered.

Compared to 1994, the consternation in the capital markets caused by the anticipation of rate hikes in 2004 was both less and shorter. In April 2004, the risk parity portfolio had a 1-month loss of –4.5% and that was it. In the next 12 month, the portfolio returned 23%.

The portfolio loss in May and June of this year is similar to that of 1994 in terms of magnitude and duration. It could be argued that the loss was entirely due to the surprise hints of QE tapering from the Fed. The surprising element was that the macroeconomic data,

TABLE 6.5 Historical Episodes of Risk Parity Losses and Subsequent Returns

	Loss (%)	**Next 12 Months**
1994 (Feb–Apr)	–10.1	15.1%
2004 (Apr)	–4.5	23.0%
2009 (Jan–Feb)	–8.7	28.0%
2013 (May–Jun)	–8.0	???

which has been indicating weak growth and tame inflation did not suggest that tapering was warranted. Therefore, the broad decline of asset prices can be attributed to the rising of risk premiums. If this is the case, historical lessons suggest that the future return could be quite positive for risk parity.

We note that Table 6.5 also lists another episode of portfolio loss for a risk parity portfolio, which occurred in January and February of 2009 when the three broad asset classes all suffered losses.* The underlying cause of this episode seemed to be driven by investors' risk aversion toward all assets rather than by any changes in the Fed's policy. Nevertheless, the portfolio similarly experienced a strong recovery in the next 12 months as risk aversion subsided and the global economy stabilized.

6.2.2 What about Bonds?

Risk parity portfolios have significant exposures to fixed-income assets simply because they have an equal risk allocation to interest rates. Thus, positive returns from fixed-income assets as well as from equities are an essential part of portfolio recovery for risk parity portfolios. This has been the case historically in all the episodes listed in Table 6.5. So how likely is it that bonds will deliver again this time? The answer is "very likely."

Compared to historical levels, UST yields are still quite low. For example, the 10-year yield is currently around 270 basis points. Compared to the risk-free rate near zero, however, the same 10-year yield actually looks quite attractive. The difference between the 10-year yield and 3-month T-bill rate—one measure of yield curve slope as shown in Figure 6.7 (solid line), has averaged about 200 basis points for the last 30 years. Today's yield curve is steeper than the historical average.

There are at least two ways to think about the steepness of the yield curve. First, from a theoretical perspective, the steepness of the yield curve indicates that the bond market might have priced in significant increases of future interest rates. If the actual rate increases are less than expected, bond returns will be positive, to say the least (see Section 4.4 in Chapter 4).

Second, from an empirical perspective, the slope of the yield curve has been a strong predictor of future excess returns for USTs. In Figure 6.7, we plot the 12-month forward excess return of the Citigroup UST index (dotted line), along with the yield curve. Graphically, the two lines seem to rise and fall together, indicating that a steepening of the yield curve tends to be followed by higher excess returns from treasury bonds.

Two quantitative measures lend support to the case. One is the correlation between the two series—it is positive 0.3. The other is the historical, conditional average excess return when the slope is around the current level. We restrict the sample to periods in which the slope was between 250 and 300 basis points. For those dates, the average excess return for the subsequent 12 months is about 2.6% and the standard deviation is about 5.2%.

It is impossible to predict precisely what bond returns will be in the future, but the steepness of the yield curve provides one indication that it is likely to be positive.

* The reason that the period of September and October 2008 is not included is because in that period, fixed-income assets actually performed quite well, albeit not well enough to offset the severe losses of risky assets.

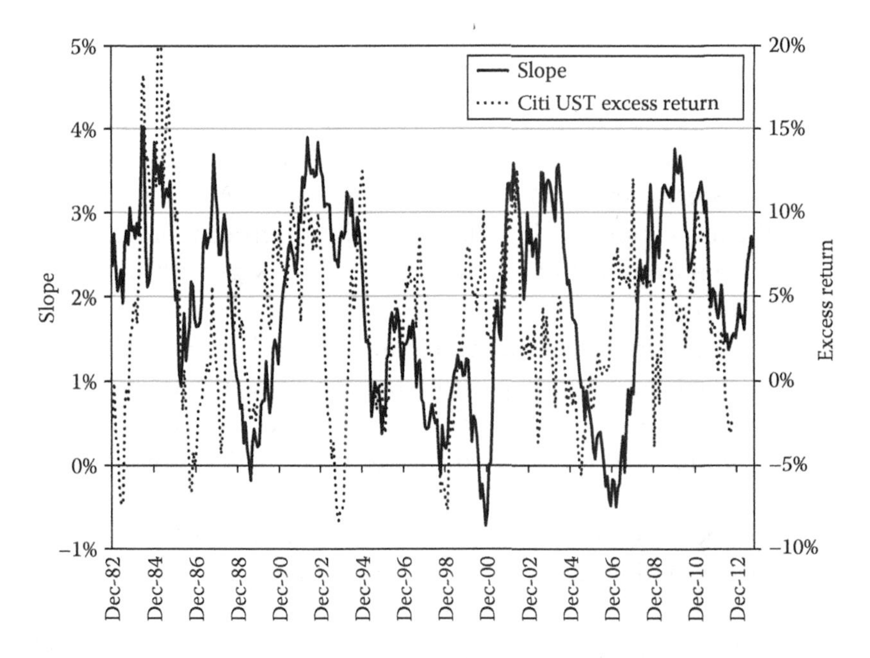

FIGURE 6.7 UST yield curve slope and 12-month forward UST excess return.

6.2.3 The Return of Diversification

From May to July, risk parity portfolios experienced heightened volatilities when broad asset classes became highly correlated. Figure 6.8 shows the 30-day rolling correlations among the daily returns of three indices: S&P 500 index, Dow Jones UBS commodity index, and Barclays US Aggregate Bond index, as well as the average of the

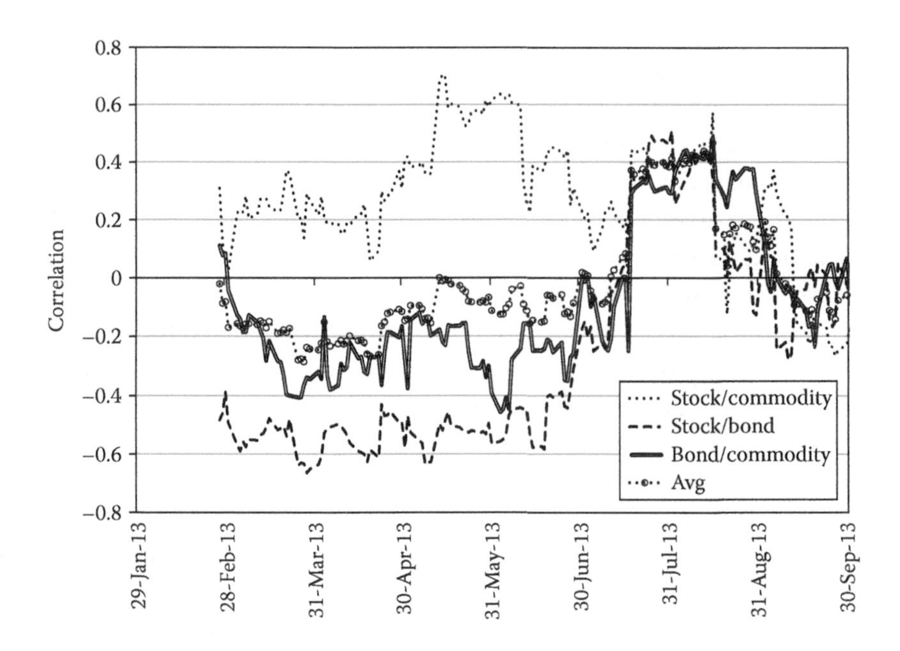

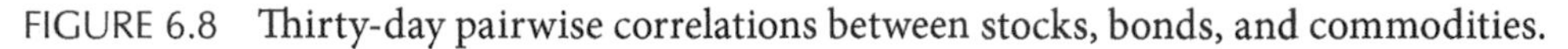
FIGURE 6.8 Thirty-day pairwise correlations between stocks, bonds, and commodities.

three pairwise correlations. From the beginning of the year to the end of May, the three pairwise correlations exhibited typical asset class relationships of the postcrisis period since 2009. There was positive correlations between risky assets, represented here by stocks and commodities, and negative correlations between risky assets and bonds, represented here as stocks and commodities versus bonds. As we wrote in a previous investment insight in Section 4.3 in Chapter 4, the macroeconomic reason for such relationships is the environment of low inflation and slow growth in the aftermath of the global financial crisis.

Starting in June, however, the correlation picture changed dramatically. All three pairwise correlations turned significantly positive and the average correlation jumped from being negative to being positive 0.4. Undoubtedly, all three assets became exposed to a single risk factor—in this case the expected change in Fed policy. Even in July when asset returns turned positive, correlations remained high. As we mentioned earlier, the most simple and perhaps the most plausible explanation is that the market was reacting to the prospect of QE tapering and demanding higher risk premiums for all asset classes by lowering their prices.

It appears that this repricing period is over. For the months of August and September, the three pairwise correlations all dropped and the average correlation has turned negative. One might think this is because the Fed has temporarily put tapering on hold. In my view, the main reason is that the market is gradually returning to pricing assets based on economic fundamentals again. In such an environment, one should expect lower correlations among different asset classes and the return of the true diversification benefits that risk parity portfolios provide.

6.2.4 Conclusion

In this research note, I present three reasons as to why the outlook has much improved for risk parity portfolios. One is based on historical evidence that rising risk premiums typically lead to higher future asset returns. It is also apparent that risk parity portfolios provide an efficient way to capture these risk premiums. The second reason is that the slope of the yield curve, as in the UST yield curve, indicates that the risk premium of interest rates is more attractive than before. This factor is suggestive of positive excess returns for the next 12 months. The third reason concerns the risk of risk parity portfolios. Risk parity portfolios experienced heightened downside risk in the months of May and June, because asset classes were repriced on the single factor of QE policy. The repricing period is likely over and asset classes will respond to macroeconomic fundamentals, resulting in lower correlations among asset classes and lower volatility for risk parity portfolios.

6.2.5 Postmortem

In the year following the tapering tantrum, the return of risk parity portfolios did recover very strongly due to the factors laid out in the investment insight above. All three risk premiums, equity, interest rate, and inflation, had significant positive returns from July 2013 to June 2014. The index returns for MSCI world, WGBI, and DJ-UBS were 19.3%, 4.8%, and

8.2%, respectively. A risk parity benchmark* had a return of 15.7%. The risk parity multi-asset portfolio managed by my team delivered a return of 20.4%.

6.3 WAITING FOR THE OTHER SHOE TO DROP†

6.3.1 QE Tapering: The First Shoe

The global financial markets had a terrible tantrum in the spring of 2013 when the former Fed chairman Bernanke hinted at the tapering of QE3—global bond yields rose sharply, followed by weakening equity and commodity prices. A year later, the US economy continues its subpar recovery, with substantial improvements in the labor market and higher albeit still subdued inflation. Meanwhile, the global equity markets have reached new highs, bond yields have stabilized, and commodity prices also recovered modestly.

It is now almost certain that the Fed will finish the purchasing of treasury bonds and mortgage-backed securities, or QE3, by October of this year. It is also widely expected that the Fed will start raising interest rate sometime in 2015, even though there is considerable uncertainty as to how the Fed would achieve the task, given its huge balance sheet of fixed-income securities and excess reserves. Nevertheless, in the next few months, the global financial markets will be anxiously waiting for the other shoe (rising interest rates) to drop. The key question is, what happens when it does?

6.3.2 Tantrum No. 2?

One could easily conjecture a replay of taper tantrum when the other shoe finally drops. In fact, this seems to be the prevailing view from Wall Street firms and many other research outlets. Specifically, the consensus appears to be that bond yields will increase sharply once more and equity markets will initially suffer again before stabilizing over the intermediate horizon. Possibly, many investors are putting their investments on hold while waiting patiently for the other shoe to drop.

In my opinion, this view is too simplistic for several reasons. First, it treats the tapering and the rate increase as two separate events. An alternative and more probable perspective is that they are two steps in one normalization process of US monetary policy. Second, one should recall that the timeframe for QE tapering was not expected when Chairman Bernanke first referenced it in May 2013. In contrast, an increase in the Fed Funds rate sometime in 2015 is widely expected hence may have already been priced in to the market. Third, and perhaps most crucially, the immediate and subsequent reactions of the global capital markets to QE tapering has been very much like reactions to interest rate increases of the past, such as the one in 2004 and to some extent the one in 1994. Even though the Fed would like to persuade (indeed they tried) investors that QE tapering amounts to less easing, not tightening, the market reaction has shown that investors have treated it as the latter and not the former. If this is the case, we need not wait until 2015 for the beginning of monetary policy tightening. It already began at the first taper in December 2013. It is then inappropriate to regard a potential rate increase in 2015 as the beginning of the tightening

* The weights are 25% in DJ-UBS index, 33% in MSCI index, and 142% in WGBI index, respectively.

† Originally written by the author in July 2014.

cycle. Therefore, historical patterns of capital markets at the beginning of tightening cycles should apply to QE tapering, not to the possible rate increase in 2015. These arguments point to the possibility that there might be no other shoe to drop.

6.3.3 Compare QE Tapering to the 2004 Rate Increases

Does QE tapering represent a tightening of monetary policy? In theory, the answer is no. If buying $85 billion of bonds is "equivalent" to cutting the Fed Funds rate by 10 bps, then buying $75 billion is the same as cutting the rate by 9 bps; it is different from raising rates. However, this kind of logic does not necessarily hold in the market place. Investors could either mistakenly think it is a tightening move or rationally regard it as a clear signal of tightening moves in the future.

No matter what the logic is, the end result is the same: the market has treated QE tapering as a tightening move by the Fed. We now qualitatively compare the market reaction to QE tapering to the market reaction to the last rate increases in 2004.*

6.3.4 UST Yields

It is obvious that Fed policy has the most immediate impact on treasury bond yields. Figure 6.9 plots the Fed Funds target rate, and treasury yields with maturities of 3 months, 2, 5, 10, and 30 years, from the end of 2003 to the end of 2004. The Fed raised the Fed Funds rate from 1% to 1.25% at the end of June and kept the pace of 25 basis point increases per meeting thereafter.

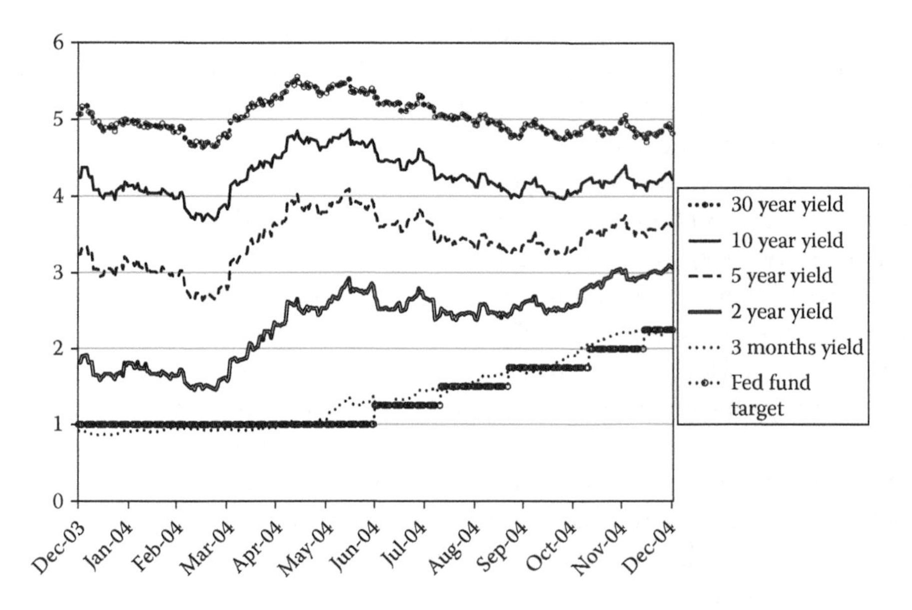

FIGURE 6.9 The Fed Fund rate and treasury yields from December 31, 2003 to December 31, 2004.

* While there are other previous episodes of rate increases, it can be argued the 2004 case is most relevant based on the openness of the Fed in its communication and the similarity in macroeconomic conditions of the US economy.

With the expectation of higher rates, bond yields started their ascent much earlier, at the beginning of March. The 2-year yield climbed by the most, followed by 5-, 10-, and 30-year yields. The 2-year yield increased by roughly 150 bps from March to the start of the actual rate increases by the Fed. After June, while the 3-month T-bill rate started to climb in tandem with the Fed Funds rate, all other yields actually declined. In the last 3 months of 2004, the 2- and 5-year yields started to increase again modestly, but the 10- and 30-year yields remained stable.

In a similar format, Figure 6.10 plots the treasury yields, together with the monthly size of bond purchases from the QE3 program, covering the end of 2012 to July 24, 2014. The axis for QE3 (in billions) is on the right and inverted.* The Fed reduced the amount of bond buying from $85 billion to $75 billion and kept the pace of $10 billion reductions at each meeting thereafter.

The movements of bond yields since 2013 mirror those between 2003 and 2004, with a few differences. First, the 5-, 10-, and 30-year bond yields rose sharply in May and June of 2013. Beginning in 2014, when QE tapering actually started, the 10- and 30-year yields declined while the 5-year yield has remained range bound. These movements are almost identical to those yield movements in Figure 6.9. The only differences are in the 2-year and 3-month yields. The 2-year yield only increased about 30 bps in the spring of 2013 and it has not moved up dramatically since. The 3-month yield has not shown any reaction to QE tapering at all.

In aggregate, it seems that the longer end of the yield curve has behaved as if QE tapering was a tightening move while the short end of yield curve is still anchored by the ZIRP.

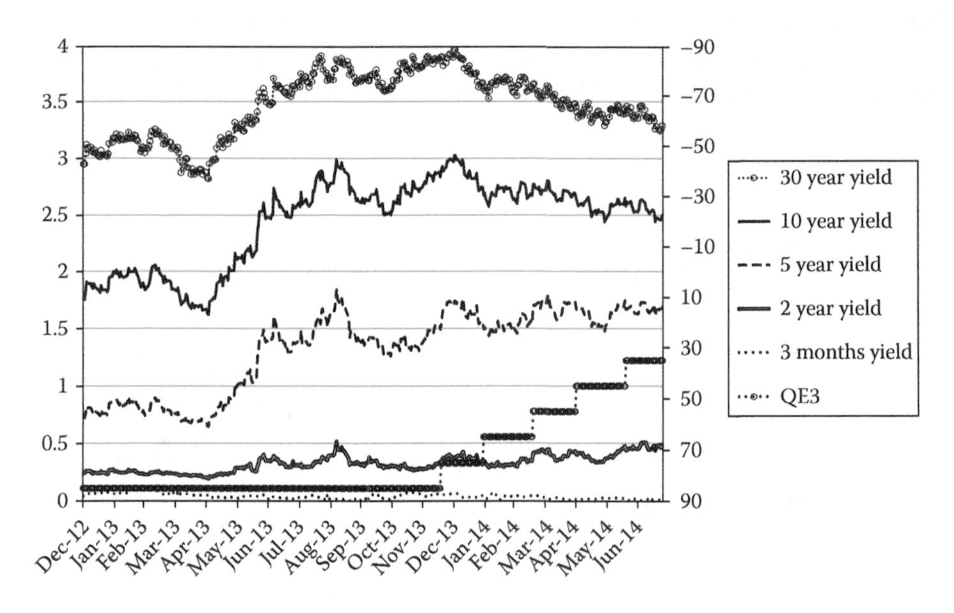

FIGURE 6.10 QE3 monthly purchase (right axis) and treasury yields from December 31, 2012 to July 24, 2014.

* The actual date of change is announcement date by the Fed.

6.3.5 Yield Curve Slopes

Another way to analyze the dynamics of bond yields is to study the relative movement of bond yields across different maturities. Yield curves typically flatten when the monetary policy is becoming tighter or less loose. Figure 6.11 plots the Fed Funds rate and the slope of treasury yield curve with two different constructions. The 10/2 yield spread was close to 250 bps at the beginning of 2004, before declining throughout the year where it ended 2004 around 100 bps.

The 30/10 spread, on the other hand, was much narrower—it was close to 100 bps in March 2004. It also declined once the rate hike was priced—declining about 40 bps by the end of 2004.

Figure 6.12 plots the 10/2 and 30/10 spreads over the recent period, together with the monthly QE purchases (right axis). The move in the 30/10 spread from about 120 bps at the beginning of the period to 80 bps at the end of the period, matches the 2004 move of the 30/10 spread extremely well. However, the 10/2 spread is rather different, due to the stickiness of the 2-year yield this time around. In other words, the 10/2 slope is almost entirely dependent on the 10-year yield. Since the beginning of 2014, this part of the curve also flattened because of the decrease in the 10-year yield. If history is any guide, we should expect it to flatten when the Fed is about to raise rates.

6.3.6 Credit Spread

We present two more comparisons between the effects of QE tapering and the effects of the 2004 tightening cycle. These two areas of focus are not directly linked to treasury markets, but they reflect the market expectations of future economic and market activities.

The first is the corporate bond market. Figure 6.13 plots the Fed Funds rate and the corporate bond OAS in 2004. The OAS tightened about 20 bps from the start of rate hikes

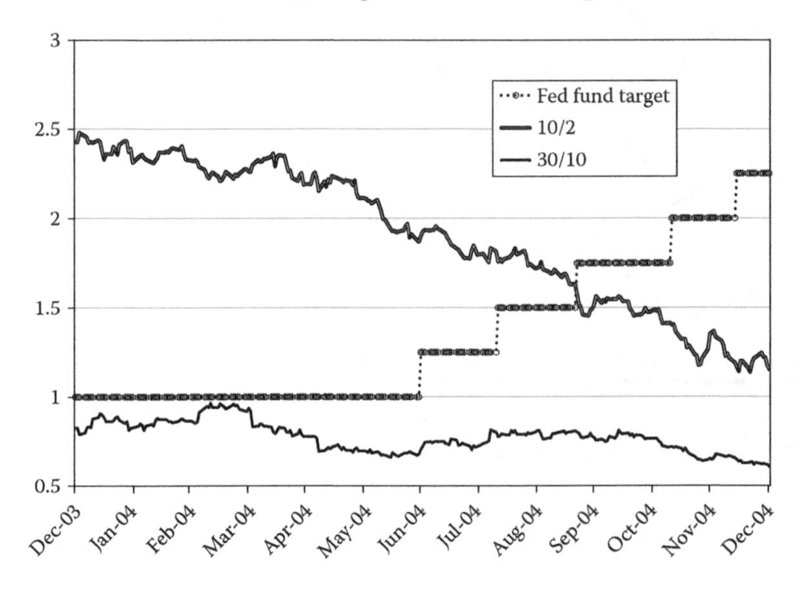

FIGURE 6.11 The Fed Fund rate and 10/2 and 30/10 yield spread from December 31, 2003 to December 31, 2004.

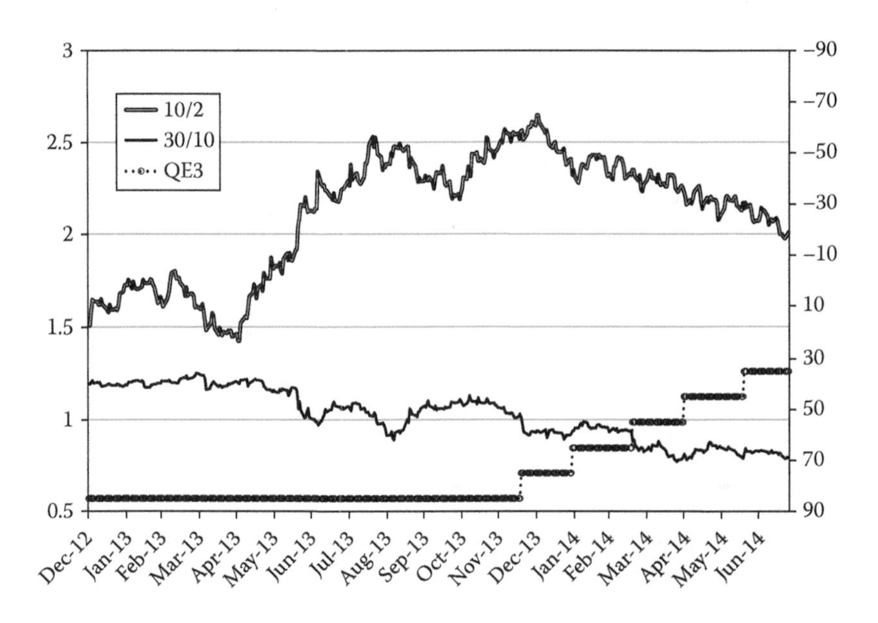

FIGURE 6.12 QE3 monthly purchase (right axis) and 10/2 and 30/10 yield spread from December 31, 2012 to July 24, 2014.

to the end of 2004, reflecting an improvement in balance sheets brought on by an improvement in the broader economy.

This picture is also true with the start of the most recent QE tapering. The OAS of the index has tightened by roughly 20 bps from January 2014 to July 2014. There was an additional 20 bps of tightening during 2013 (Figure 6.14).

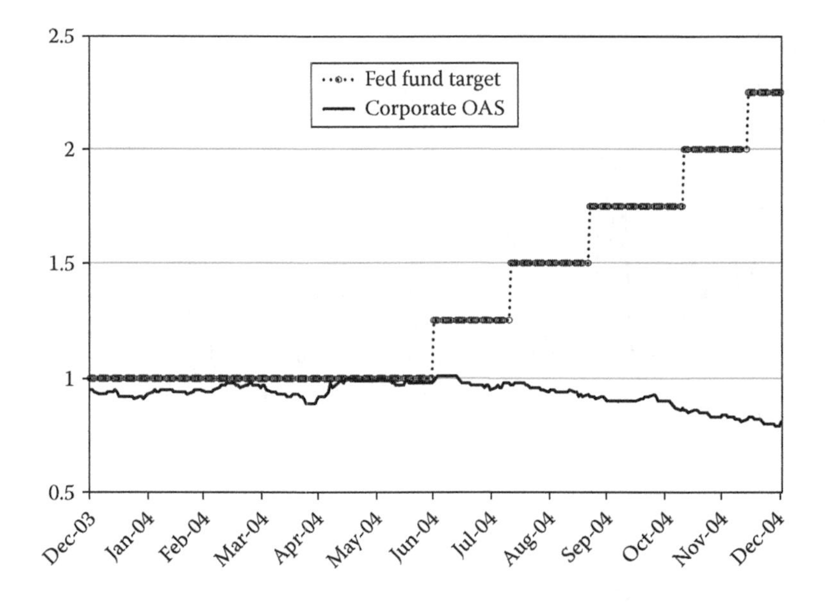

FIGURE 6.13 The Fed Fund rate and corporate bond option-adjusted spread from December 31, 2003 to December 31, 2004.

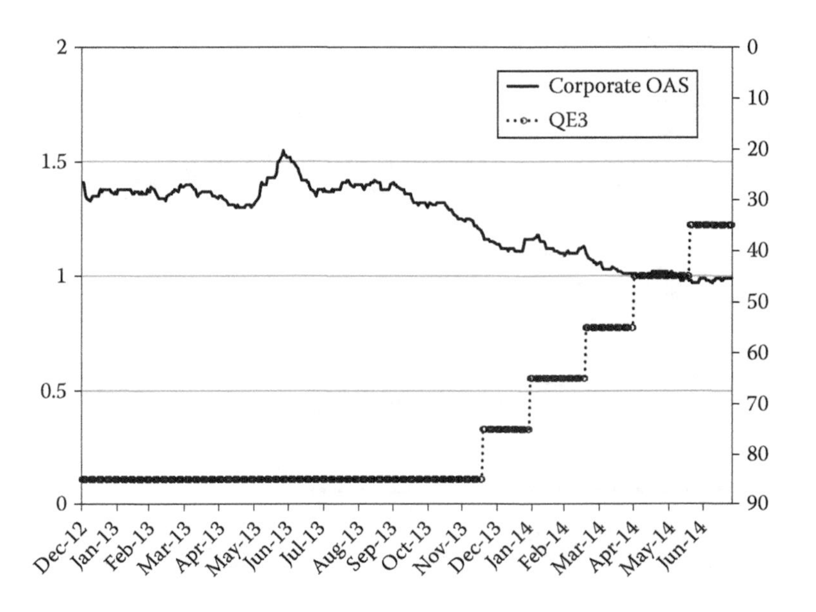

FIGURE 6.14 QE3 monthly purchase (right axis) and corporate bond OAS from December 31, 2012 to July 24, 2014.

6.3.7 Market Volatility

Many investors have been surprised by the low volatility of many markets during recent months. This phenomenon of low volatility has happened in the past when the Fed started raising rates. Figure 6.15 plots the Fed Funds rate and the VIX index in 2004. Following the initial rate increases, the equity implied volatility declined from an average of 17% to

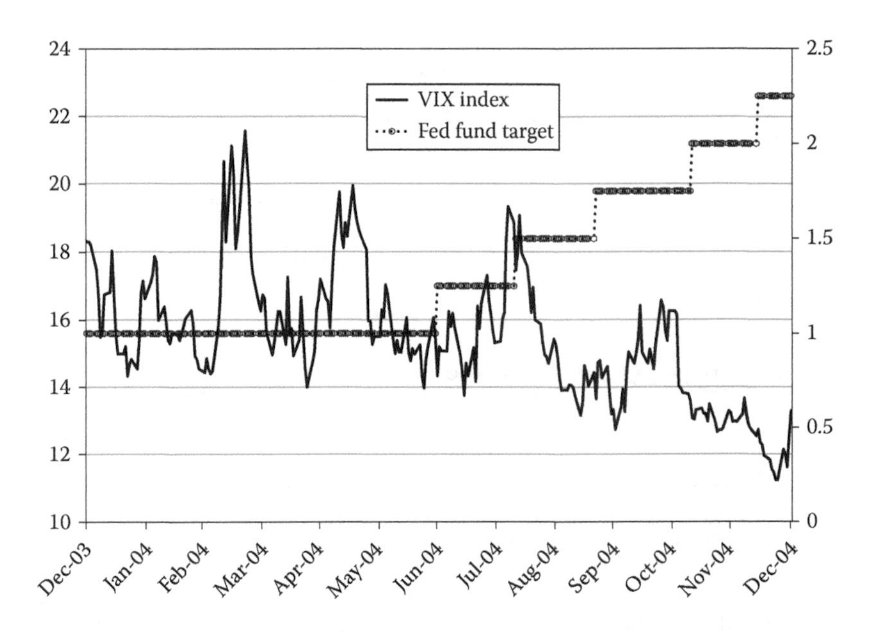

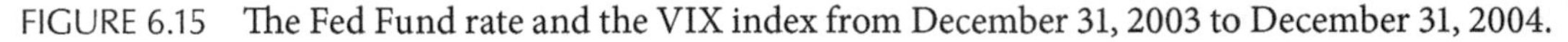

FIGURE 6.15 The Fed Fund rate and the VIX index from December 31, 2003 to December 31, 2004.

an average of 14%. Perhaps, this reflects investors' confidence in both the economy and the equity market. Alternatively, it might be because the uncertainty of monetary policy was resolved.

It appears that this is happening again in the equity market now. Figure 6.16 plots the size of the QE3 program's monthly purchases along with the level of the VIX index. Since March, equity implied volatility has collapsed from around 15% to 12%. The realized volatility is perhaps even lower, despite a multitude of geopolitical risks from the Middle East and the Russia/Ukraine border.

The low volatility phenomenon has not been restricted to just the equity market. The implied volatility of treasury bonds has also been low.

The MOVE index is a weighted average of yield volatility across different maturities. As we can see from Figure 6.17, during 2004, it reached a peak a few months before the beginning of the Fed tightening only to decline once the tightening actually began. The picture is similar for the MOVE index and the current QE tapering. The former peaked during May and June of 2013 and it has been declining ever since. This is another piece of evidence that the bond market has been treating the tapering of QE as if it were the start of Fed tightening (Figure 6.18).

6.3.8 Conclusion

In order to combat the global financial crisis and deal with its aftermath, the Fed adopted a ZIRP in 2008 and soon after embarked on a mixture of unconventional monetary policy programs, including QE1, QE2, Operation Twist, and QE3. Since the beginning of 2014, the Fed has been working toward winding down QE3 and it is expected to drop the other shoe (raise short-term interest rates) sometime in 2015.

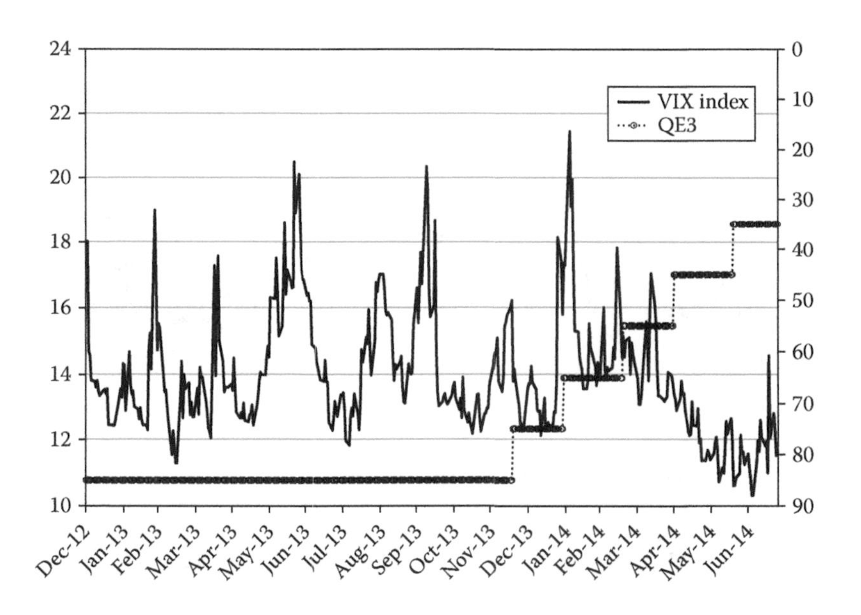

FIGURE 6.16 QE3 monthly purchase (right axis) and the VIX index from December 31, 2012 to July 24, 2014.

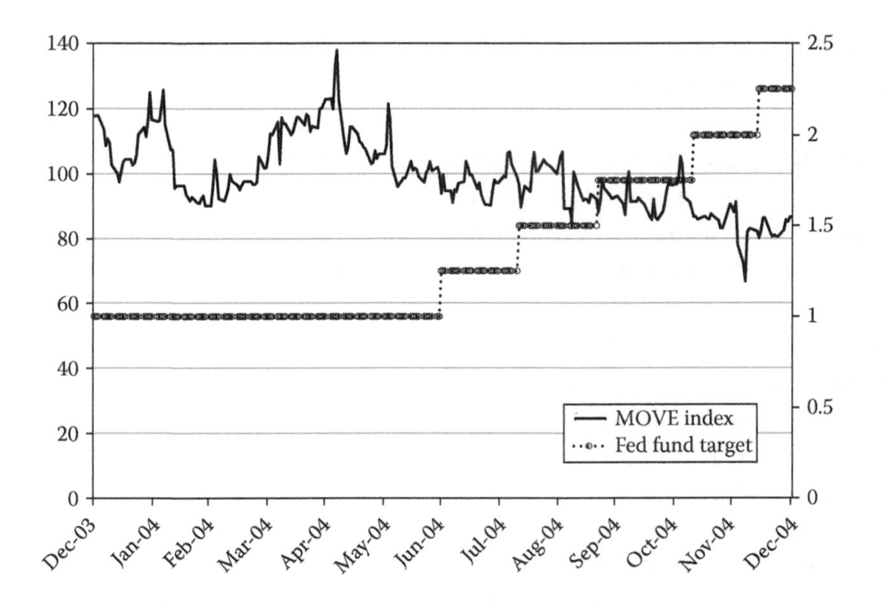

FIGURE 6.17 The Fed Fund rate (right axis) and the MOVE index from December 31, 2003 to December 31, 2004.

While many strategists expected a replay of the taper tantrum when that happens, I have presented several arguments as to why this may not be the case. There is strong evidence that the market has also reacted to QE tapering as it were the beginning of a tightening cycle. Therefore, when the Fed eventually raises the short-term rate as a continuing step toward the normalization of monetary policy, the market reaction could be much muted, except for perhaps the short end of the yield curve.

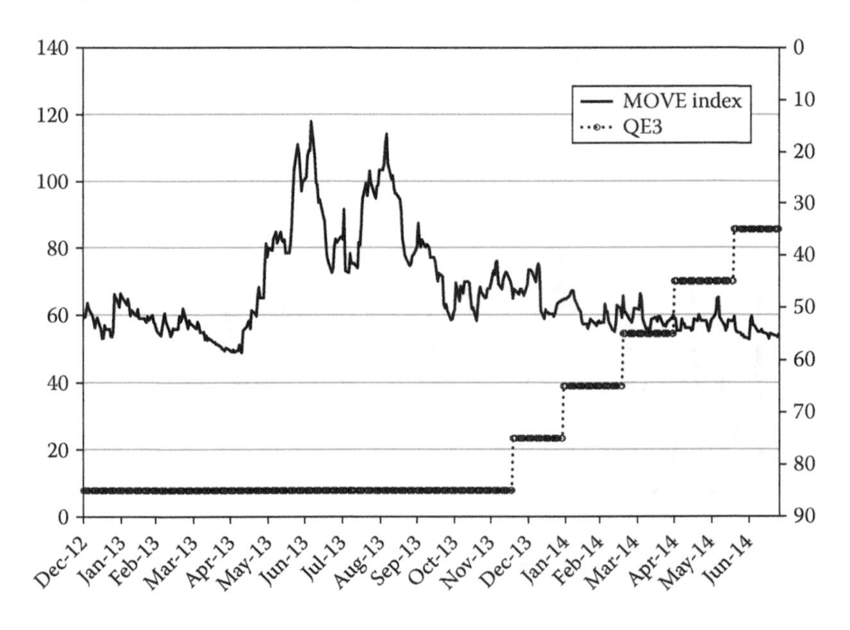

FIGURE 6.18 QE3 monthly purchase (right axis) and the MOVE index from December 31, 2012 to July 24, 2014.

6.3.9 Postmortem

As of October 2015, the Fed has yet to raise the short-term rate more than a year after the taper tantrum. While the job market has improved, the inflation remained low. The Fed is debating whether or not to raise the Fed Fund rate in 2015. While the bond yields in the United States have increased in 2015, the magnitude of increase has been rather modest compared to taper tantrum. For example, the 10-year UST yield has gone from 2.17% to 2.40%. The 10-year yields of other government bonds have shown similar increases. The WGBI bond index had a return of −0.68%. Meanwhile, the equity markets have shown moderate appreciation. The MSCI World Index had a return of 2.92% in the first half of 2015. The commodity market, measured by the DJUBS commodity index, is slightly down in the first half of 2015 with a return of −1.56%. There has been no sign of the other shoe dropping.

6.4 IS US BECOMING JAPAN?*

Only with extraordinary monetary and fiscal support, did the US economy exit the great recession in the middle of 2009 and start on the path of recovery. But what a rocky path it has been! Barely 1 year later, the recovery ran into a soft patch in the middle of 2010, which prompted the Fed to embark on another round of quantitative easing (QE2). As US economic growth sputtered once again in the first half of this year, global equity markets sold off sharply and UST bond yields tumbled to new lows in the month of August. There is now a general concern that the United States may follow the footsteps of Japan, which has experienced two lost decades after its real estate and equity bubbles burst at the end of the 1980s. How real is this concern?

6.4.1 United States Is Already Half Way There

By many well-known measures, the United States is already half way down the path to Japan with its first lost decade after the technology bubble burst in 2000. The first measure is the equity market. From December 2000 to December 2010, the S&P 500 index delivered an annualized return of 1.2%, underperforming both T-bills and treasury bonds. The next measure is the economy. The average economic growth during this period has been less than 2%, substantially less than previous decades. The two economic recoveries, one after the tech bubble and the other after the recent credit bubble were subpar with minimal job creation compared to previous recoveries. Today, the unemployment rate stands above 9% and the options to accelerate growth and job creation seem limited.

There is another measure that is perhaps less known but, in my view, characterizes this lost decade. It is the negative correlation between stock and bond returns. Figure 6.19 plots the cumulative growth of the S&P 500 index and the Citigroup UST index (left axis) from 1979 as well as the 3-year rolling correlation between the monthly returns of the two indices (right axis). Prior to 2000, the correlation was consistently positive even though it dipped lower after the 1987 stock market crash and during the recession of 1991–1992.

* Originally written by the author in September 2011.

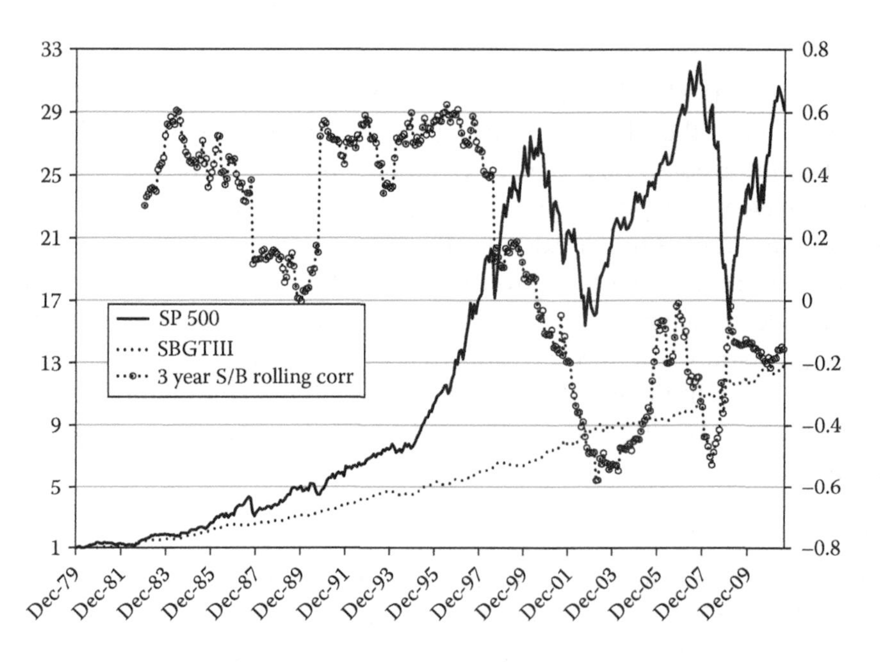

FIGURE 6.19 Cumulative returns of the S&P 500 index and the Citigroup UST index plus the 3-year rolling correlation between the two indices.

After 2000, however, the correlation has been consistently negative ever since. What is the cause of this negative correlation?

6.4.2 Living in a Low Growth and Low Inflation Environment

I think this persistent negative correlation is a symptom of asset markets in a low growth and low inflation environment. In addition to slow growth in the last decade, the United States also had subdued inflation with core CPI around 2% and headline CPI near 2.5%. One can attribute the low level of inflation to both internal (the Fed policy on price stability since 1980s) and external forces (manufacturing growth in emerging markets). Stocks and bonds typically respond to both inflation and growth shocks, but in opposite ways. For example, when inflation declines, both stocks and bonds respond favorably, causing a positive correlation between the two. However, when growth declines, stocks tend to suffer and bond prices rise, resulting in a negative correlation. When inflation was rising and then falling rapidly in the 1970s and 1980s, stocks and bonds had positive correlation because the effect of inflation was dominating. However, in a low inflation environment, the effect of growth dominates. In other words, asset markets become more sensitive to growth causing the correlation between stocks and bonds to become decidedly more negative.

Japan, which suffered two decades of low inflation and slow growth, had an early start in seeing this negative correlation. Since 1991, the average real GDP growth in Japan has been close to 1% while average inflation has been close to zero.

Figure 6.20 plots the cumulative growth of the MSCI Japan index and the WGBI Japan index (left axis) from 1988 and the 3-year rolling correlation between the monthly returns of the two indices (right axis). Approximately 2 years after the equity bubble burst, the

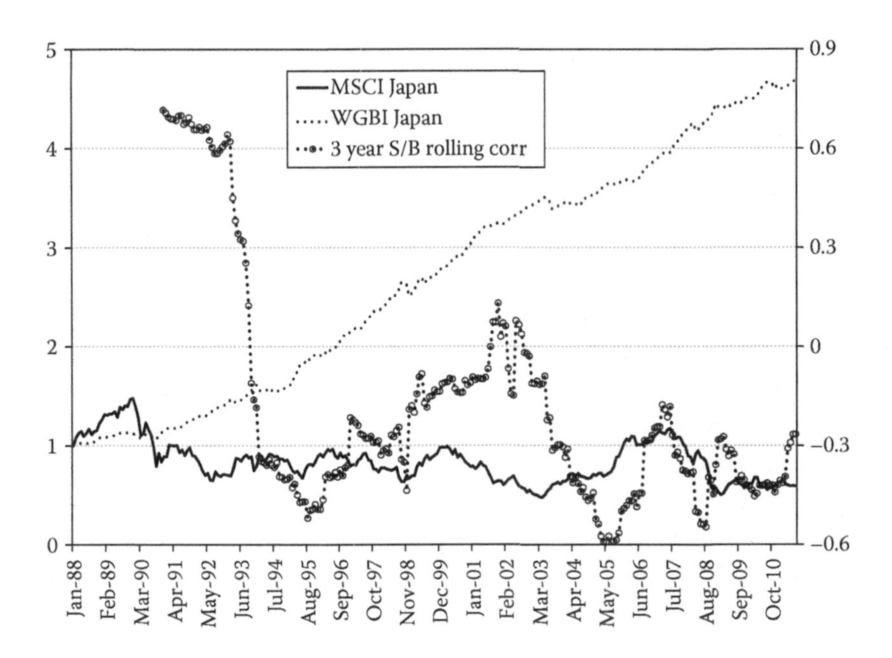

FIGURE 6.20 Cumulative returns of the MSCI Japan index and the WGBI Japan index plus the 3-year rolling correlation between the two indices.

correlation turned negative and it has stayed negative ever since except for a brief period in 2001 and 2002. At this point, it is hard to dispute that the United States is remarkably similar to Japan in terms of the macroeconomic environment and the behavior of asset markets.

6.4.3 Asset Allocation Implications

This negative correlation is likely to continue. We almost see this on a daily basis with risk on or risk off markets that seems to care exclusively about growth and little about inflation or deficits. Over the long term, if the United States does follow Japan with another lost decade, bonds would likely perform well while equities would likely continue to languish with high volatility. Today, the consequence of another recession would be dire and this is why both equity and bond markets are reacting very violently to that possibility. On the other hand, there are many fundamental reasons that suggest the United States might turn out differently. For instance, the US policy response to the credit crisis has been much more forceful than what was seen in Japan. Fed Chairman Ben Bernanke, being a scholar of the Great Depression and the Japanese economic crisis, has acted more aggressively in terms of cutting the short-term interest rate and expanding the Fed's balance sheet to lower long-term interest rates. Another fundamental reason might be demographics. While Japan's population growth has been anemic, the US's population has grown steadily near 1% over the last 30 years. If (a big if) these and other factors could turn back the tide, and lead to stronger growth in the United States, equities would likely perform well and bond yields would probably move higher.

Hence, a negative correlation between stocks and bonds is consistent with the expected divergence between the two asset classes' returns. Proper diversification becomes imperative in reducing return volatility as well as generating returns. If anything, there is a bright side to the negative correlation—it reduces portfolio volatility quite significantly. For instance, consider a 60/40 portfolio, assuming stock volatility of 18% and bond volatility of 6%. If the correlation is positive 0.3 then the portfolio volatility is 11.8%. On the other hand, if the correlation is negative 0.3 then the volatility drops to 10.3%.

Even though the volatility is lower, a 60/40 portfolio cannot be expected to generate much return if the Japan scenario is replayed in the United States. The reason is simple—a 60/40 portfolio is not properly diversified at all on the risk of economic growth since it has about 95% of its risk allocation to equity. A risk parity portfolio, which allocates risk equally between stocks and bonds, provides true diversification along the growth dimension. It is therefore a much better solution to take advantage of the negative correlation and generate positive returns.

Consider again the case of Japan. A simple back test can illustrate the difference. As Table 6.6 shows, from the perspective of a Japanese investor, a 60/40 portfolio based on MSCI Japan and WGBI Japan has volatility of 10.62% and a return of merely 0.98% from 1996 to July 2011. On the other hand, a risk parity portfolio with 45% in equity and 135% in bonds would have a volatility of 8.23% and a return of 6.83%.

6.4.4 Conclusion

The possibility of the United States following Japan to another lost decade is real and dangerous, given the low inflation and low growth environment we find ourselves in after the credit bubble burst. A traditional 60/40 portfolio would perform poorly if that happens. Investors should prepare for that possibility by diversifying the growth risk in their portfolios. A risk parity portfolio provides such diversification. As we have shown in the Japan example, it can generate decent returns with lower risks.

6.4.5 Postmortem

As of October 2015, it seems that the United States has not followed Japan in repeating the experience of deflation and slow growth. Partly due to the extraordinary effort of the Fed, led by former Chairman Bernarke, an expert in depression economics, who was quick to apply lessons learned from the Great Depression and Japan's deflation, the US economy has shown great improvement in employment and economic growth. However, inflation remains low and below the target of 2%, and growth continues to be subpar compared to other periods of economic recovery and market expectations.

The best candidate for repeating the Japan experience turns out to be the Euro zone, which has been mired in disinflation and minimum growth. Currently, the Euro zone's core CPI is just 0.6% and sovereign bond yields of many Euro zone countries have fallen

TABLE 6.6 Return and Risk of 60/40 and Risk Parity Japan Portfolios

	60/40 Japan (%)	Risk Parity Japan (%)
Return	0.98	6.83
Risk	10.62	8.23

to levels comparable to JGB yields. For example, the German 10-year bond yield is near 50 basis points. In hindsight, there are many possible reasons for the Euro-zone dilemma, including slow monetary stimulus from the ECB, austerity by governments of many countries, and a single currency. Nevertheless, few investors expected this to happen a few years ago. It remains to be seen whether the Euro zone could escape the deflation trap with the ECB's own QE program.

6.5 RISK PARITY AND INFLATION*

6.5.1 Risk Parity and Inflation Protection

Risk parity portfolios balance risk contribution from high-risk assets such as equity and commodity and low-risk assets such as IG bonds. To achieve true diversification as well as return/risk target, it is often necessary to have high asset exposure, resulting in leverage at the portfolio level. Compared to traditional asset allocation, the bond exposure is higher because IG bonds have lower return volatility.

With bond yields near the lower end of the historical range, one natural concern about risk parity portfolios is the bond position. What if higher inflation returns? In the near term, in our view, the economic environment of high unemployment, low capacity utilization, and declining credit demand is not conducive to higher inflation, even with a record level of government stimulus. Higher inflation is certainly possible in the long run. For one thing, we have been in a disinflationary world for the last three decades, the trend might not continue. Even though forecasting long-term asset returns is always difficult if not impossible, it is prudent to study the investment consequence of a higher inflation scenario.

Rising inflation is not good for nominal bonds as it leads to higher nominal interest rates. It will not be kind to equity investment either. In theory, equities should provide inflation hedging since equity investors have a claim to real company earnings. But this has not been the case in practice as the experience in the 1970s shows. What would happen to risk parity portfolios when higher inflation returns?

Two aspects of a risk parity portfolio shall provide defense against inflation risk and generate real rates of returns that are lacking in traditional asset allocation portfolios. The first line of defense is through exposure to commodities and inflation-linked bonds. A baseline risk allocation of 20% to these two asset classes amounts to a significant exposure of 40% on average. The second is a dynamic risk allocation process, which allows us to increase inflation hedging if it becomes necessary. Our research has shown dynamic risk allocation is likely to capture inflation cycles thus providing additional inflation hedging in a potentially higher inflation regime.

6.5.2 Historical Analysis

To further study the issue, we consider a historical perspective by examining risk parity portfolios in the high inflation period of the 1970s. Figure 6.21 shows annual CPI in the

* Originally written by the author in March 2010.

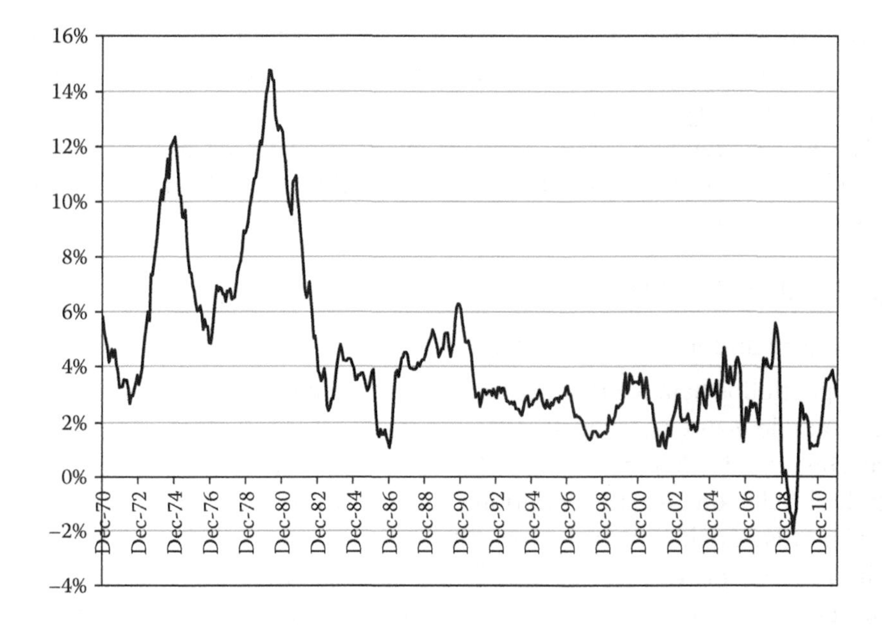

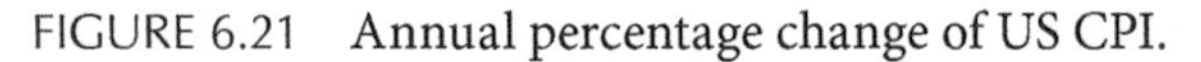
FIGURE 6.21 Annual percentage change of US CPI.

United States from 1970 to 2009. The 1970s are characterized by generally higher inflation that averaged 7.4% and two oil shocks that caused inflation to spike above 10%. The following three decades saw much lower inflation—CPI averaged 5.4% in the period from 1980 to 1989, 2.9% from 1990 to 1999, and 2.5% from 2000 to 2009.

How would have risk parity performed in the first decade? To answer the question we simulate its performance using four asset classes: US equity (S&P 500), US nominal bonds (5 year UST bonds), commodities (GSCI), and US TIPS. The selection of these four was based on data availability and their relevance in risk parity portfolios.* The inclusion of TIPS is crucial in any inflation study. However, the return data are limiting—for example, the US government only started issuing TIPS in 1997. For this study, we simulate US TIPS returns using prevailing real yields, derived from nominal bond yield and long-term inflation.

Table 6.7 shows four decades of the annualized returns of the four asset classes as well as cash returns and CPI. Cash returns are important since risk parity portfolios depend partly on excess returns of risky assets over cash. In the 1970s when inflation is high and volatile, cash, nominal bonds and stocks had positive returns but they all underperformed

TABLE 6.7 Annualized Percentage Change in CPI and Annualized Returns of Financial Assets

	CPI (%)	Cash (%)	GSCI (%)	TIPS (%)	US Bond (%)	S&P 500 (%)
1970–1979	7.4	6.5	21.2	10.4	6.7	5.9
1980–1989	5.1	9.1	10.7	6.6	12.1	17.6
1990–1999	2.9	5.0	3.9	5.3	7.2	18.2
2000–2009	2.5	2.8	5.1	5.7	5.3	–3.5

* In reality, risk parity portfolios invest in more than four-asset classes.

CPI, resulting in negative real returns. GSCI is by far the best-performing asset while TIPS also returned 3% higher than CPI. The fortunes reversed in the following two decades. With declining inflation, came booming stock market and sizable nominal bond returns while the return of real assets declined. But both GSCI and TIPS still had positive real returns. Disinflation continued in the last decade, GSCI, TIPS, and nominal bonds all had low single-digit returns beating cash and CPI. In contrast, the stock market had negative returns for the last 10 years with the bursting of the tech and credit bubbles.

We simulate returns of a risk parity portfolio based on these four asset classes (five if cash is included), according to the same exact investment process we follow in practice. We assign risk allocation targets to each asset and estimate a covariance matrix; we then find the portfolio that matches the risk allocation targets and total portfolio risk for the next period. In addition, we simulated two versions of risk parity portfolios: one with fixed risk targets and the other with dynamic risk allocation targets based on our proprietary model. For comparison, we also show returns of a 60/40 portfolio using the same asset returns.

Table 6.8 shows the returns of the risk parity foundation (RPF), risk parity dynamic (RPD), and 60/40 portfolios for the last four decades. Figure 6.22 is a graphical illustration and the combined results. There are several notable results in this table. First, the risk parity portfolios performed quite well in the 1970s. While the 60/40 portfolio returned 6.5%, which is below the CPI rate of 7.4%, the risk parity portfolios had returns above 10% and real returns above 3%. Exposure to commodities and TIPS in the risk parity portfolio was able to provide inflation hedging, generating positive real rates of return. Second, we note that risk parity with dynamic risk allocation had higher returns than the foundation version in all four decades, with outperformance ranging from 0.6% to 1.3%. This shows that dynamic risk allocation could add value in different economic environments. Third, risk parity generated similar performance to 60/40 in the 1980s and 1990s. This is achieved with a much lower allocation to equities when the equity market had probably the best 20 years in the history of investing.* Fourth, even though risk parity's returns in the 2000s are lower than the previous decades, the real returns were still positive 3% compared to the negative real return of the 60/40 portfolio. Finally, we note that the returns of risk parity portfolios are more stable over the four periods, due to their balanced risk allocation and true diversification. On the contrary, the returns of the 60/40 portfolios depend mostly on that of equity with poor diversification and high variability over the four periods.

TABLE 6.8 Simulated Returns for Risk Parity Portfolios and 60/40 Portfolio

	CPI (%)	Risk Parity Foundation (%)	Risk Parity Dynamic (%)	60/40 (%)
1970–1979	7.4	10.5	11.7	6.5
1980–1989	5.1	14.3	14.9	15.7
1990–1999	2.9	12.2	13.5	13.9
2000–2009	2.5	4.3	5.2	0.4

* Risk parity portfolio with more asset classes would likely have outperformed 60/40 in those 20 years due to greater diversification.

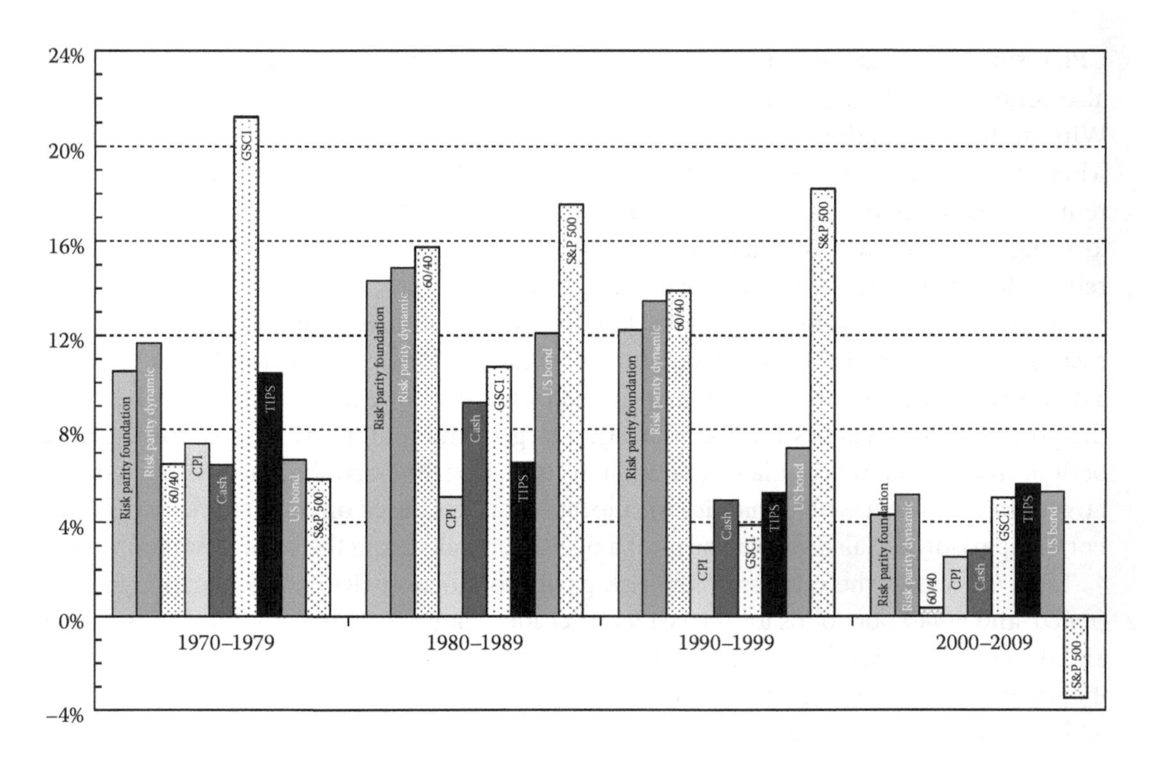

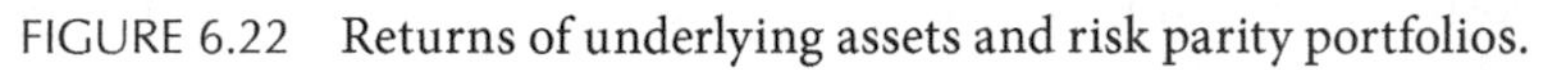
FIGURE 6.22 Returns of underlying assets and risk parity portfolios.

6.5.3 A Closer Look at the 1970s

While the 1970s had in general higher average inflation, the level of inflation was also quite volatile. Therefore, it is worth a detailed look of asset returns from year to year to see its impact. Table 6.9 expands the returns for the decade to individual calendar years. Consistent with Figure 6.21, there was two inflation cycles in the 10-year period. From 1970 to 1972, inflation declined; from 1973 to 1974, we had the first inflation spike; from 1975 to 1976, inflation declined again; and from 1977 to 1979, inflation spiked the second time. So it was down-up–down-up. During the two inflation spikes, asset classes behaved

TABLE 6.9 Annual Portfolio and Asset Returns

	RPF (%)	RPD (%)	60/40 (%)	CPI (%)	Cash (%)	GSCI (%)	TIPS (%)	US Bond (%)	S&P 500 (%)
1970	21.2	21.2	9.8	5.7	6.5	15.1	22.1	18.2	3.9
1971	17.5	16.3	12.1	2.9	4.4	21.1	10.7	8.3	14.3
1972	17.6	18.4	12.3	3.8	4.2	42.4	4.3	2.8	19.0
1973	10.5	14.9	–7.1	8.7	7.3	75.0	12.4	4.5	–14.7
1974	–3.1	1.6	–14.3	12.2	8.0	39.5	17.6	5.7	–26.5
1975	14.9	11.0	24.9	6.8	5.9	–17.2	8.4	7.4	37.2
1976	19.7	21.5	20.0	4.9	5.1	–11.9	14.1	13.8	23.9
1977	–3.9	–2.9	–3.9	6.7	5.5	10.4	2.1	1.1	–7.2
1978	3.5	5.8	4.6	9.1	7.6	31.6	2.8	1.0	6.6
1979	10.4	11.8	13.2	13.3	10.6	33.8	11.0	5.3	18.6

more or less with a predictable pattern. In 1973 and 1974, equities declined sharply while commodity and TIPS kept up with rising inflation and nominal bonds had positive nominal returns but negative real returns. From 1977 to 1979, commodities performed well throughout while other asset classes showed some deviations. Equities had two meager years but managed to come back in 1979. Nominal bonds had very low returns; TIPS outperformed nominal bonds but failed to beat inflation due to the widening of the real yield. Overall, risk parity portfolios outperformed 60/40 portfolios in those periods while suffering minimal losses in absolute terms, due to inflation hedging and dynamic risk allocation.

6.5.4 Conclusion

Rising inflation is potentially harmful to nominal asset returns. Substantial allocation to real assets such as those in risk parity portfolios is necessary for inflation hedging and generating real returns. Using simulation, we are able to examine risk parity portfolios during the high inflation period of 1970s. There are four main findings. First, risk parity portfolios were able to provide a real return (close to 4%) while 60/40 failed to do so in the 1970s. Second, a dynamic risk allocation process was adaptive in all periods, providing additional defense against rising inflation. Third, a closer examination of annual data from 1970 to 1979 confirmed the general finding. Finally, while the focus of the chapter is on inflation risk, the results again show that risk parity portfolios, with a balanced risk allocation to a variety of asset classes—high risk and low risk, nominal, and real, deliver true diversification and more stable long-term returns than traditional asset allocation portfolios.

For this section, we omit any postmortem analysis. As of October 2015, there does not seem to be any sign of significantly higher inflation in the horizon.

CHAPTER 7

The "Wild West" of Risk Parity

RISK PARITY STRATEGIES, DUE to its sound investment principles and proven track record in recent years, have gained a certain degree of acceptance in the investment community, despite much initial criticism. Many asset managers have started offering and managing risk parity products on behalf of investors. In addition, many capable institutional investors carved out some of their own assets to run internal risk parity portfolios, based on their own conceptual understanding and sometimes with the help of external managers. It appears that there is a proliferation of risk parity strategies.

How does one know whether these risk parity strategies are actually managed according to a risk parity principle? This is an important and reasonable question to ask by investors and prospects, for several reasons. First, there is no uniformly accepted interpretation of risk parity. Second, there is no common benchmark for risk parity. These ambiguities have both pros and cons. One of the benefits is that there is no "closet indexer." All managers and investors have the freedom and indeed incentive to research and implement different ideas based on their own understanding and insights. These efforts might lead to diversification among different managers. However, one drawback is that some of the interpretations and ideas might be wrong and detrimental to performance. In the pursuit of innovation, it is inevitable that some ideas might be wrong and they do not work in practice. Indeed, there has been wide return differences among many risk parity managers in recent years, due to the idiosyncrasy of individual managers. This has shattered an old belief that all risk parity portfolios are similar.

This situation is not unique to risk parity. The problem is much pronounced in hedge fund space. As we discussed previously, there is neither a common definition nor common benchmark for hedge funds. Many hedge fund managers use discretionary and fundamental judgments in their investment process, which makes it extremely hard to measure their performance against any prescribed benchmarks.

The situation for risk parity managers is more manageable. The commonalities among different risk parity strategies are numerous: returns are all from various risk premiums, all portfolios are long-only, and many risk parity products target risk levels around that of a traditional 60/40 portfolios through portfolio leverage.

The potential differences are also numerous, however. A partial list might include the following and each item on the list would lead to some questions about risk parity portfolios.

1. Selection and "classification" of asset classes: Do managers classify asset classes in the correct premium buckets* and do the selected asset classes represent a balanced mix of various risk premiums?
2. Strategic risk allocation: Do managers allocate risk to independent risk premium or do they double up on two correlated risks like equity and credit? Do they have equal risk allocation to independent risk premiums or do they have an intentional overweight to one risk premium at the expense of others?† What are the rationales for such an overweight?
3. Asset class exposures: How managers implement asset exposures makes the difference. Do they simply invest in capitalization-weighted indices or do they implement bottom-up risk parity and to what degree?‡
4. Tactical shifts: Managers have the latitude to make tactical shifts in many forms. The size and manner of the shifts can be very different and impactful. Do they make moderate shifts or drastic shifts that might undermine risk parity principle?§ Are these shifts systematic or discretionary?
5. Tail risk hedging: After the 2008 global financial crisis, many managers added tail-risk hedging in their risk parity portfolios.¶ In terms of costs and timing, is this akin to buying insurance after a hurricane? Do these insurance schemes work over the long run? Do they have any investment value besides providing a psychological "comfort"?

This partial list and associated questions does not even begin to describe the "wild west" of risk parity! In this chapter, we provide three investment insights that aim to provide some answers to certain questions related to the top-down allocation decisions of risk parity managers. In the first insight, we reiterate the definition of risk parity in terms of a balanced risk allocation to three primary risk premiums. We then employ a return-based style analysis, which had been developed for analyzing traditional asset allocation and equity funds, to derive the effective asset weights and risk allocations for several risk parity managers. These effective weights are remarkably accurate in modeling actual returns of risk parity managers, both in sample and out-of-sample. Furthermore, based on the risk decomposition of these effective asset weights, it is found that some risk parity managers have a systematic bias in their strategic risk allocation. In the second insight, we carry out a follow-up out-of-sample study by applying the results of the style analysis to risk parity

* Some mistakenly put high yield and emerging market bonds into interest rate risk even though they represent mostly equity risk.

† Some target equity risk at 50%, while interest rate and inflation risks make up the remaining 50%.

‡ Some used just a handful of commodity contracts for commodity exposure. Some implement interest rate exposures with call options, potentially eroding interest rate premium.

§ Some made seismic shifts on rare occasions while other might regularly make large tactical asset allocation decisions.

¶ Some essentially have 60/40 portfolios along with CDX and variance swaps and falsely market the strategy as risk parity.

performance in 2013. It is found that the model performance matched managers' actual performance rather well.

The third investment insight studies the efficacy of a stop-loss policy, which is one form of tail risk hedging, for asset classes and risk parity portfolios. We report that stop-loss works reasonably well for commodities. However, it is generally ineffective for risk parity portfolios.

7.1 ARE RISK PARITY MANGERS RISK PARITY?*

In recent years, risk parity managers are popping up, as the Chinese saying goes, like bamboo shoots after a spring rain. In the case of risk parity though, "the spring rain" was the "tsunami" of the global financial crisis, which delivered devastating losses to traditional capital-based asset allocation portfolios and prompted a significant increase in investors' interest in risk-based portfolios such as risk parity.

But how do we know these risk parity products are truly risk parity, not something else merely dressed up, or other different products but misclassified by consultants and database providers? This is an important question for investors because an "in-name-only" risk parity portfolio will fail to provide the desired stable investment returns as well as risk diversification expected of risk parity strategies.

This is not an easy question. In order to get an answer to the question, we have to address at least two issues. First, there is no consensus on the definition and interpretation of the risk parity principle. Second, there is no simple way for the investing public to tell whether a given manager adheres to the principle. But these difficulties shall not deter us from the search for answers. In this research note, I shall aim to define the principle of risk parity investing and then examine several risk parity managers quantitatively using a return-based style analysis. In light of the right definition of risk parity and the results of the style analysis, we shall conclude that not all "risk parity" managers are risk parity.

7.1.1 What Is Risk Parity?

As someone who initiated risk parity research, and in 2005 coined the term "risk parity," I would like to share with readers my perspectives about risk parity as an investment approach. The best way to do this is to start with what is not risk parity.

First, we can all agree that a 60/40 stock/bond portfolio is not risk parity, because the so-called balanced (in capital) portfolio is terribly unbalanced in risk allocation with between 90% and 95% of the portfolio risk in stocks. Second, a portfolio with an equal risk allocation to all select asset classes is not necessarily risk parity. Imagine a manager has chosen four equity asset classes and one fixed-income asset class in an asset allocation portfolio. An equal risk contribution portfolio would have 80% risk in stocks and 20% in bonds (Chaves et al., 2011), which is perhaps an improvement over the 60/40 portfolio but definitely not risk parity. Conversely, if there were four fixed-income assets and one equity asset class, then an equal risk contribution portfolio is not risk parity either.

* Originally written by the author in November 2012.

This example thus highlights the importance of parity along the right dimension. Obviously, it is not the number of assets. Is it the asset category like stocks and bonds? This brings me to the third point: a portfolio with equal risk allocation to select asset categories is not necessarily risk parity. Suppose the manager above has realized his mistake and decides to include four fixed-income asset classes to balance out the number of equity assets. However, the four fixed-income asset classes are HY, EM debt, inflation-linked bonds, and IG bonds. Is an equal risk contribution portfolio from these eight asset classes risk parity? It certainly is not, because these four fixed-income asset classes all have varying degrees of equity risk or inflation risk exposure. In the case of HY, it is almost all equity (See Chapter 2, Section 2.1).

So what is risk parity? The key word in risk parity is risk. A risk parity portfolio, at a minimum, must have a balanced risk allocation along the economic risk dimensions that have a major impact on the portfolio's returns. For risk parity asset allocation portfolios, the key risk dimensions are growth and inflation risk.

Associated with these risks, there are risk premiums that are provided by different asset classes. Along the growth risk dimension, there are equity risk premium and interest rate premium and along the inflation risk dimension, there are real return premium and nominal return premium. When we piece these premiums together with the balanced exposure to growth and inflation risks, then it is evident (see detailed discussion in Chapter 2) that a risk parity portfolio should have a balanced risk contribution from three sources: (1) equity risk; (2) interest rate risk; and (3) inflation risk.

Some asset classes fit into these three risks directly. Stocks are equity risk, government bonds are mostly interest rate risk, and commodities are inflation risk. Other "hybrid" asset classes, such as the four fixed-income asset classes mentioned above, can be broken down to the three risks by either qualitative or quantitative analyses.

7.1.2 A Style Analysis of Risk Parity

I define risk parity as portfolios that have balanced risk contribution from equity, interest rate, and inflation. I now examine seven risk parity managers listed in the eVestment database by performing a return-based style analysis and then mapping their effective asset allocation mixes to risk allocations.

Return-based style analysis was introduced by Sharpe (1988, 1992) to analyze asset managers in asset allocation and equity portfolios. The original technique is designed for long-only unlevered portfolios; I have extended it to analyze long-only leveraged portfolios. Next I select the sample period and return indices.

Our sample period covers the trailing 3 years from October 2009 to September 2012, during which monthly returns for all seven managers are available.* While it is possible to stretch the sample period further back, many risk parity managers have evolved their strategies in one way or the other after the global financial crisis of 2008. Therefore, the data from the last 3 years might be a better representation of their current style.

* One of the managers stopped reporting in June 2012, though.

The choice of return indices also warrants some consideration. On the one hand, a choice of too few might not give a decent coverage of all the managers' investment choices. On the other hand, too many indices might lead to multi-collinearity and overfitting. However, I have found that while the effective asset mix might change with the choice of the return indices, the final risk allocation to the three risk sources is rather robust. This is because changes of the effective asset mix as we add more assets mainly happen within assets that have high correlations and therefore similar types of risks. I thus opt to use a sufficient but not exhaustive number of return indices for a good fit of the style analysis.

Table 7.1 shows the effective asset mixes for the seven risk parity managers from A to G. There are 12 asset classes included in the analysis. For commodities or inflation risk, I use the DJ-UBS commodity index. Among the fixed-income asset classes, UST, MBS, and WGBI ex US are almost all interest rate risk; Credit and EM Debts are a combination of interest rate risk and equity risk; TIPS is a combination of interest rate risk and inflation risk; and HY is all equity risk. Finally, there are four equity asset classes representing equity risk.

I make several remarks about the results in Table 7.1. First, the style fit is quite good. With the exception of manager D, the R-squared is either close to or above 90%. Second, the leverage ratios are between 200% and 300% except for manager F whose leverage is 188%. Manager B has the highest leverage at 269%. But I caution that this leverage comparison is not indicative of the level of portfolio risk, since 93% of manager B's portfolio is in the WGBI ex US index whose return volatility is very low whereas manager F has an exposure of 65% in four equity asset classes and HY combined.

Furthermore, as we shall see in the next section, the fact that, a portfolio, such as those of manager F, is levered and has substantial notional exposures to fixed-income assets does not necessarily prove the portfolio is risk parity. In the discussion of risk parity portfolios, the use of leverage to balance risk allocation is often a hotly debated issue, which in my view is misguided. But it is crucial to point out that portfolio leverage is necessary for risk

TABLE 7.1 Effective Asset Mixes of Seven Risk Parity Managers, and Total Leverages and R-Squared of the Fit

	A (%)	B (%)	C (%)	D (%)	E (%)	F (%)	G (%)
DJUBS	18	7	15	25	21	13	17
BarCap UST	0	43	66	38	65	17	0
WGBI ex US	87	93	57	93	52	9	74
BarCap MBS	0	0	0	0	0	18	7
Citi US TIP	57	79	46	29	55	50	37
BarCap Credit	30	18	7	0	0	15	0
Citi EM Debt	0	7	0	1	0	0	14
BarCap US HY	1	0	4	0	0	8	32
S&P 500	0	0	8	15	4	37	2
MSCI ex US	22	11	3	0	28	19	19
R2000	13	9	15	5	8	0	15
MSCI EM	0	2	5	0	0	1	9
Leverage	228	269	226	205	232	188	225
R-squared	92	94	95	82	89	96	96

parity (to achieve a certain risk level) but leverage by itself is not a sufficient condition for an asset allocation portfolio to be risk parity.

Third, among all asset classes, managers have common exposure to some but not to others. The common exposures seem to include commodities, global sovereign bonds, inflation linked bonds, global developed equities, and small cap equities. But since many asset classes are highly correlated one must take these asset exposure with caution.

7.1.3 Risk Allocations of Risk Parity Managers

Given the effective asset mixes in Table 7.1, we can now derive the risk allocation to the individual asset classes, using a covariance matrix of asset returns. Most risk parity managers use various quantitative methods and long-term historical returns to compute the covariance matrix. A covariance matrix based on the last 3 years of returns, which is used in the style analysis, is too short term and susceptible to distortion introduced by the particular macroeconomic environment. I thus use monthly returns of a much longer period to calculate the covariance matrix used for deriving risk allocation.

Once I have the risk allocation of the individual asset classes, I combine them into an aggregated risk allocation to the three risk sources: equity, interest rate, and inflation and the results are shown in Figure 7.1.

The question now is whether these risk allocations are balanced. It is obvious that manager F and G do not pass the test since both have a very low-risk allocation to interest rate risk. Manager G, in particular, has a risk profile that is very similar to that of the 60/40 portfolio, except some of the equity risk is now diverted to inflation risk. In Table 7.2, I also aggregate equity and inflation risk into "risk on" risk and relabel interest risk as "risk off"

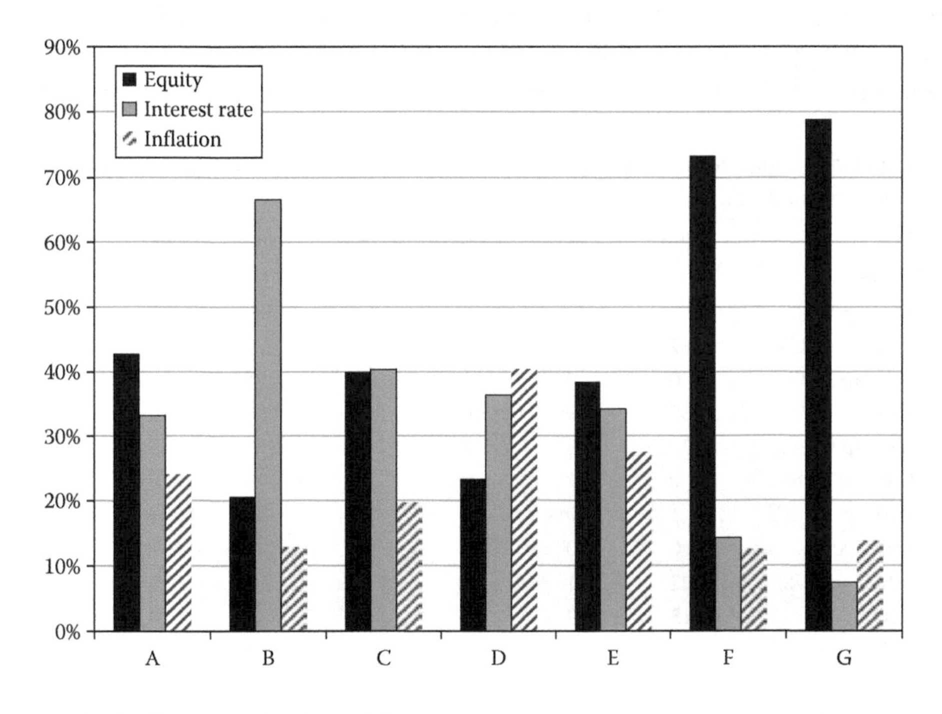

FIGURE 7.1 Risk allocations to three risk types.

TABLE 7.2 Risk On Combines Risks in Equity and Inflation and Risk Off is the Interest Rate Risk

	A (%)	B (%)	C (%)	D (%)	E (%)	F (%)	G (%)
Risk on	67	33	60	64	66	86	93
Risk off	33	67	40	36	34	14	7

risk to reflect the recent risk on risk off market phenomenon whereby many risky assets including equities and commodities move in tandem. In this perspective, the 93/7 split between risk on and risk off makes manager G no different from a 60/40 manager.

The risk profile of manager F is similar, with only 14% in interest rate risk and 86% in equity and inflation risk combined. It is apparent that both manager F and G are heavily exposed to equity or growth risk, with no meaningful difference to the traditional 60/40 portfolios. Yet, both managers have significant notional exposure to fixed income and significant portfolio leverages. How can this be? There are three reasons. First, the interest rate exposures are in low risk asset classes. Second, many fixed-income exposures have embedded equity risk. Third, they have high notional weights in equities. Due to these reasons, appeared as they might be, managers F and G are not risk parity.

Manager B is different from risk parity in the opposite way—its interest rate exposure accounts for 67% or two-thirds of the risk budget while equity and inflation risks account for only 21% and 13%, respectively. The reason is its effective asset allocation mix has a low weight in equity assets (22%) and commodities (7%) and very high weights in both nominal and inflation-linked bonds. The aggregate risk on risk is only 33%. It appears that manager B is invested in not risk parity but a highly levered portfolio with mostly fixed income plus some exposures to equity and real assets in commodities and inflation-linked bonds.

Figure 7.1 shows that the rest of the four managers: A, C, D, and E, all have a rather balanced risk allocation to three types of risks—they all fall in the range from 20% to 40% in each risk. While this group shares more similarity in their risk profiles, according to Table 7.2, manager A appears the most growth oriented with 67% of its risk in risk on and manager C appears the least with 60% of its risk there.

7.1.4 Return "Tests"

The style analysis and the breakdown of the risk allocation to equity, interest rate, and inflation risk might appear abstract. But their practical implication is clear: managers with higher allocations to equity and inflation risks would perform well in risk on markets but do poorly in risk off markets. In contrast, managers with higher allocations to interest rate risk would perform well in risk off markets but lag in risk on markets. It is especially true when the markets are volatile, whether being an either up or down market.

The past 3 years provide many test cases for our prediction. We choose December 2010 as the risk on case, where risky assets rallied strongly and government bond yields went up significantly. For the risk off case, we choose August and September of 2011, where risky assets suffered severe losses and UST yields hit new lows. Based on our analysis, there should be strong correlation between managers' allocation to risk on risk and their returns in December 2010.

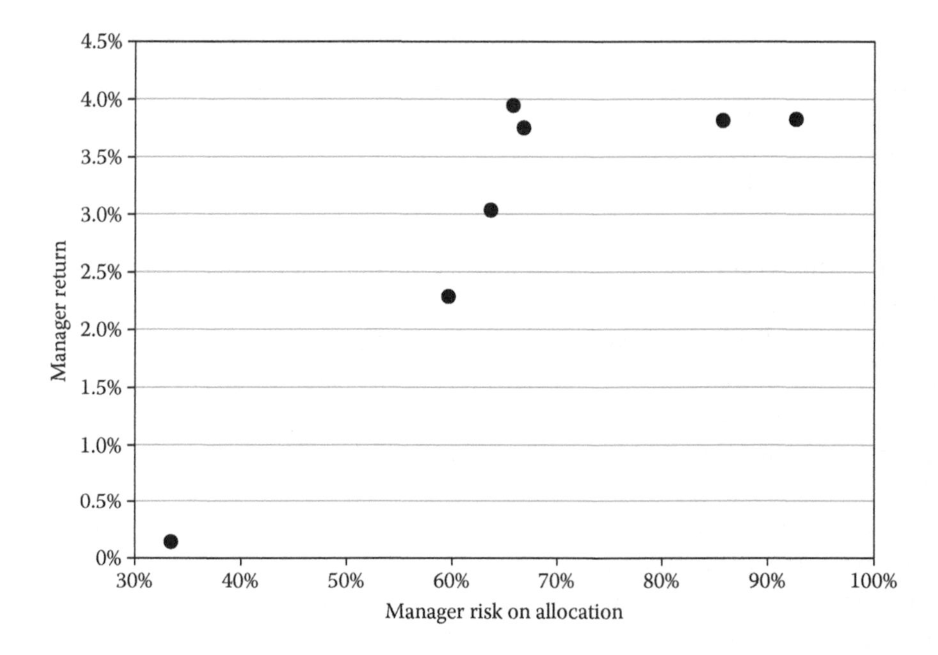

FIGURE 7.2 Managers' risk allocation to risk on assets and their performance in December 2010.

Figure 7.2 plots risk allocations to equity and inflation versus their returns in December 2010. Indeed, there is strong correlation between them. Manager B with the least amount of risk on risk has the lowest return while managers F and G with the highest risk on risk attain the highest returns.

For the case of risk off markets in August and September of 2011, Figure 7.3 plots the risk off allocations versus managers' returns. Indeed, managers with a very low risk allocation to risk off assets suffered the most losses while the manager with the highest risk allocation to risk off assets only had minimal losses. These "event" studies provide strong validation to our return-based style analysis and risk allocation decomposition to three risk sources.

7.1.5 Conclusion

Risk parity as an alternative asset allocation approach differs from a traditional capital-based approach by balancing the risk allocation from various sources. Because its implementation has considerable freedom and the concept of risk parity is open to different interpretations, it is hard for investors to tell the difference between various products in the market place.

In this investment insight, I argue that a true risk parity portfolio should have balanced risk exposures to the economic risks of growth and inflation, and as a consequence, balanced, not necessarily equal, risk contribution from three sources of risks: equity, interest rate, and inflation.

Measured against this criterion, we examine seven "risk parity" managers in the eVestment database with return-based style analysis and find that at least three managers have investment styles that are significantly different from risk parity. Two of those managers have dominant equity or growth risk exposures that are reminiscent of the traditional

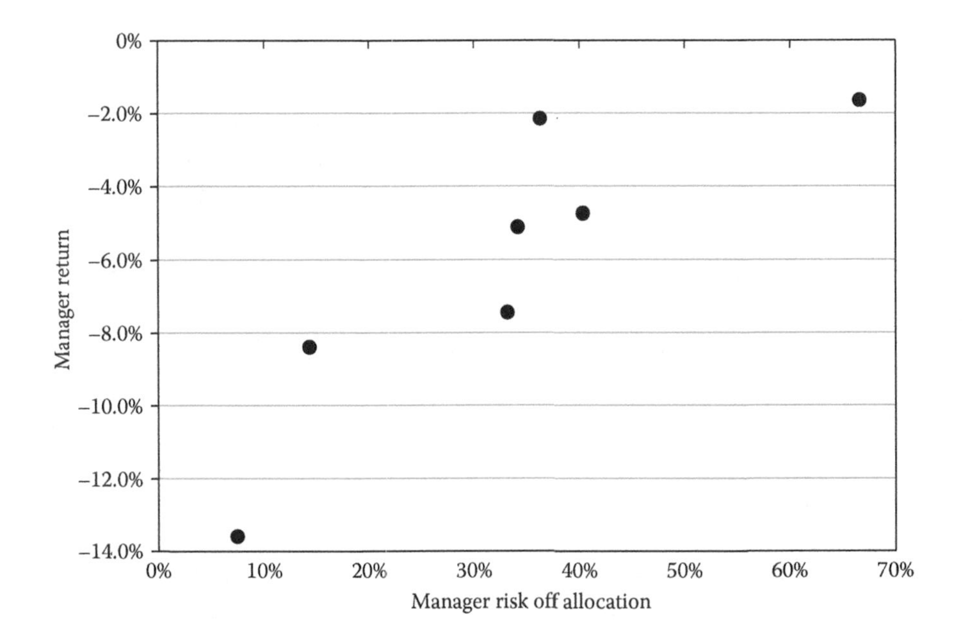

FIGURE 7.3 Managers' risk allocation to risk off (interest rate risk) and their performance in August and September 2011.

60/40 portfolios and the other manager has concentrated interest rate exposure with much less equity risk than what risk parity would imply.

Given these results, it is interesting to ponder why some "risk parity" managers are not risk parity. One possible reason is some trivial but consequential misunderstanding of the risk parity principle as discussed previously—either parity in the number of asset classes or parity in the category of assets could lead to unintended risk concentration across risk dimensions that are based on economic fundamentals.

Another possibility is that some managers are intentionally making a strategic decision to overweigh significantly one type of risk over other types of risks. For example, a common but so far mistaken prediction over the last 3 years was the rise of interest rates. Maybe, some "risk parity" managers have made this view a dominant theme of their strategies. But even if this were the case, the magnitude of their active decisions, inferred from their risk concentration in both equity and inflation risks, seems extraordinary. On the contrary, the manager with concentration in interest rate risk could be expressing a strategic preference over quality fixed-income assets over growth assets.

Finally, one probably should not underestimate the commercial force of a popular investment strategy, which has drawn many enthusiastic entrants, who nevertheless have a misconceived notion of the risk parity concept.

All the reasons above mean that it is important for investors to conduct a thorough analysis of risk parity portfolios beyond asset weights or leverage ratios. Only a deeper understanding of how risk parity portfolios are implemented, in terms of the risk dimensions, strategic, and tactical risk decisions will determine whether they are actually risk parity.

7.2 PREDICTING RISK PARITY MANAGERS' PERFORMANCE*

In the preceding investment insight, I presented a style analysis of seven risk parity managers and derived each manager's effective asset weights based on their monthly returns from the eVestment database as well as a set of traditional asset index returns. At the time, I found an excellent in-sample fit with an average R-square above 90%. Another startling discovery from the analysis was that three out of the seven managers were not truly at risk parity because their risk allocations to the three primary risks (equity, interest rate, and inflation) were not as balanced as implied by the risk parity approach. On the surface, all managers appeared to be risk parity, with significant notional weights in fixed-income assets and substantial portfolio leverage. A more detailed risk analysis revealed that two managers held strong equity biases due to their concentration in growth assets (stocks and low-grade bonds) and one manager had strong interest rate bias due to its concentration in high-grade bonds.

A year has since passed since the original analysis. A natural question is, how accurate is the style analysis out-of-sample? In this follow-up note, I evaluate the prediction of the style analysis against the actual performance of the universe of risk parity managers and find strong agreement between the two. First, on both an individual and aggregated basis, the return predictions are quite accurate. Second, in a period when equity risk had strong positive rewards while interest rate risk delivered weak or negative returns, the risk analysis correctly identifies the underperformance of the manager who had a concentrated interest rate risk allocation.

The analysis presented in this research note indicates that the effective asset mixes are reasonably accurate in predicting future returns. If so, they could serve, to some extent, as a proxy for a strategic benchmark for risk parity managers. The real-time performance during the out-of-sample period, albeit a short one, provides investors with an opportunity to evaluate their performance against these proxy benchmarks.

7.2.1 Summary Results of the Style Analysis

For ease of reference, we summarize the results of the previous section here. Table 7.3 shows the result of the effective asset mixes from the previous style analysis.† In addition, Table 7.3 also provides the average asset weights. On average, these six managers have the following exposures: 16% in commodities, represented by the DJ-UBS index; 174% in bonds, most of which is in USTs (38%), non-US government bonds (65%), and US TIPS (53%), as well as 34% in equities, represented by US and non-US developed market stocks and US small-cap stocks. The average allocation to MBS, EM debt, EM equity, and HY bonds is very low. The average leverage is 225% and the average R-squared is 91%.

While the style analysis has excellent overall fit, we should not expect that it is accurate in all asset class weights. For example, I strongly suspect that the effective weights in EM

* Originally written by the author in October 2013.

† One of the managers (manager G in the original study) had stopped reporting; thus we exclude it from the current analysis.

TABLE 7.3 Effective Asset Mixes of Six Risk Parity Managers, Total Leverage, and R-Squared of the Fit

	A (%)	B (%)	C (%)	D (%)	E (%)	F (%)	AVG (%)
DJUBS	18	7	15	25	21	13	16
BarCap UST	0	43	66	38	65	17	38
WGBI ex US	87	93	57	93	52	9	65
BarCap MBS	0	0	0	0	0	18	3
Citi US TIP	57	79	46	29	55	50	53
BarCap Credit	30	18	7	0	0	15	12
Citi EM Debt	0	7	0	1	0	0	1
BarCap US HY	1	0	4	0	0	8	2
S&P 500	0	0	8	15	4	37	11
MSCI ex US	22	11	3	0	28	19	14
R2000	13	9	15	5	8	0	8
MSCI EM	0	2	5	0	0	1	1
Leverage	**228**	**269**	**226**	**205**	**232**	**188**	**225**
R-squared	**92**	**94**	**95**	**82**	**89**	**96**	**91**

debt and EM equity are lower than the actual weights taken by managers. However, these shortfalls are probably offset by weights in other equity asset classes offering the same risk exposure. The crucial question is the accuracy of the return forecasts based on these effective asset weights.

From the perspective of risk analysis, we group risk allocations, derived from the style weights, into equity, interest rate, and inflation risks. The results are shown in Figure 7.4. On average, the risk allocations to both equity and interest rate risks are 40% and 37%,

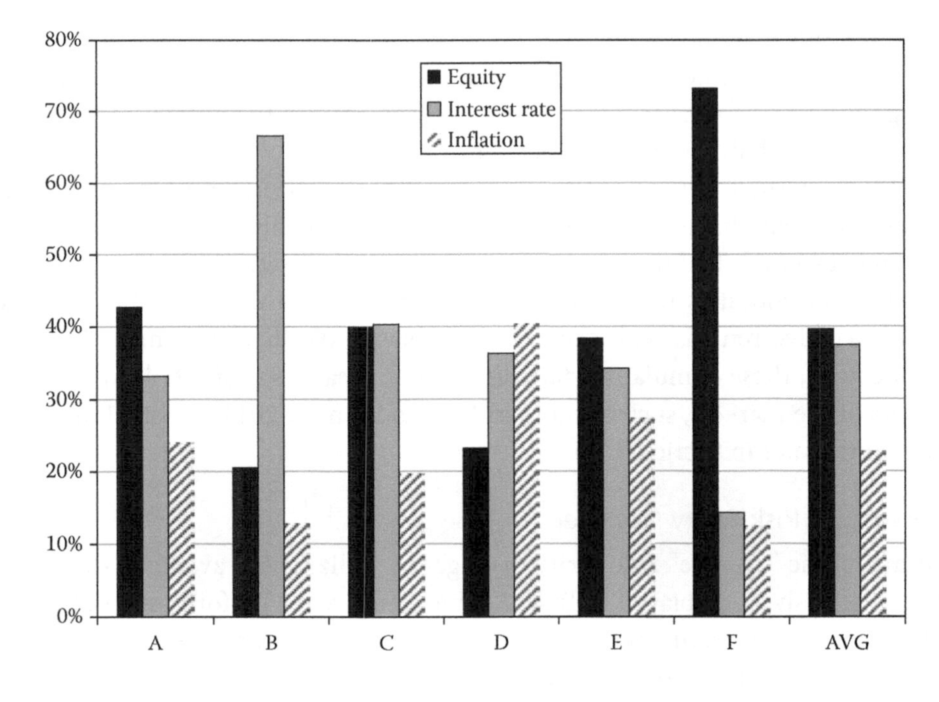

FIGURE 7.4 Risk allocations to three risk types.

respectively, while the risk allocation to inflation risk is around 23%. These risk exposures are in general balanced since a risk parity approach does not mean they have to be equal. Two managers have very different risk profiles, however. Manager B shows significant risk concentration in interest rate risk while manager F shows significant risk concentration in equity risk. Consequently, we would expect the returns of these two managers to deviate from the rest and specifically from each other, depending on different market environments. Will this be true out-of-sample?

7.2.2 Out-of-Sample Market Environment

Our style analysis essentially provides models for the six risk parity managers. These models are estimated over 3 years of return data ending in September 2012. The out-of-sample period covers 1 year, beginning in October 2012 and ending in September 2013. While this subsequent period was not great for risk parity strategies in general, one could hardly ask for a better market environment as far as for testing asset allocation models. Moreover, the behavior of asset returns during this period is remarkably different from the behavior of asset returns during the 3-year in-sample period.

First, this period is marked by sharp movements in bond yields. Government bond yields reached historically low levels in the early months of 2013 and then rose sharply in May and June of 2013. Since risk parity managers typically have a substantial allocation to interest rates, the interest rate volatility over this period was particularly impactful to their performance. How would the models from the style analysis track the actual performance of the managers in this interest rate environment?

Second, May and June of 2013 saw large drawdowns in risk parity portfolios because most assets or risks delivered negative returns. It is natural to examine how our predictions panned out in this extreme market environment.

Third, returns from different asset classes deviated strongly from each other, with equity delivering the best performance. This would likely accentuate the differences between risk parity managers with different risk allocations.

Table 7.4 shows the 1-year cumulative returns, the annualized monthly return standard deviations, and their Sharpe ratios of the asset indices over this period. Developed equity markets and HY bonds had the best results in terms of both return and Sharpe ratio. Real assets such as commodities and inflation-linked bonds had the worst results. Fixed-income assets had negative returns with the exception of the World Government Bond ex US Index. Of course, these cumulative statistics do not reveal some of the dramatic monthly movements of the markets, such as those in May and June of 2013. We shall discuss these specific months later in Section 7.2.5.

7.2.3 "Average" Risk Parity Manager

We can model the "average" risk parity manager by utilizing the average asset allocation from the style analysis, displayed in Table 7.3. We compare the performance of this portfolio to the average of the actual returns from October 2012 to September 2013.

Figure 7.5 displays the two return streams, together with the error of the forecasts. We note that the forecast model accurately predicts the average returns of managers throughout

TABLE 7.4 Summary of Returns, Risk, and Sharpe Ratios from October 2012 and September 2013

	Return (%)	Standard Deviation (%)	Sharpe
DJUBS	−14.4	32.5	−0.44
BarCap UST	−2.2	9.4	−0.23
WGBI ex US	2.1	9.5	0.22
BarCap MBS	−1.3	8.8	−0.14
Citi US TIP	−7.0	22.0	−0.32
BarCap Credit	−2.0	16.6	−0.12
Citi EM Debt	−4.2	29.9	−0.14
BarCap US HY	7.1	15.5	0.46
S&P 500	19.3	31.3	0.62
MSCI ex US	22.2	30.9	0.72
R2000	30.0	41.8	0.72
MSCI EM	0.9	40.6	0.02

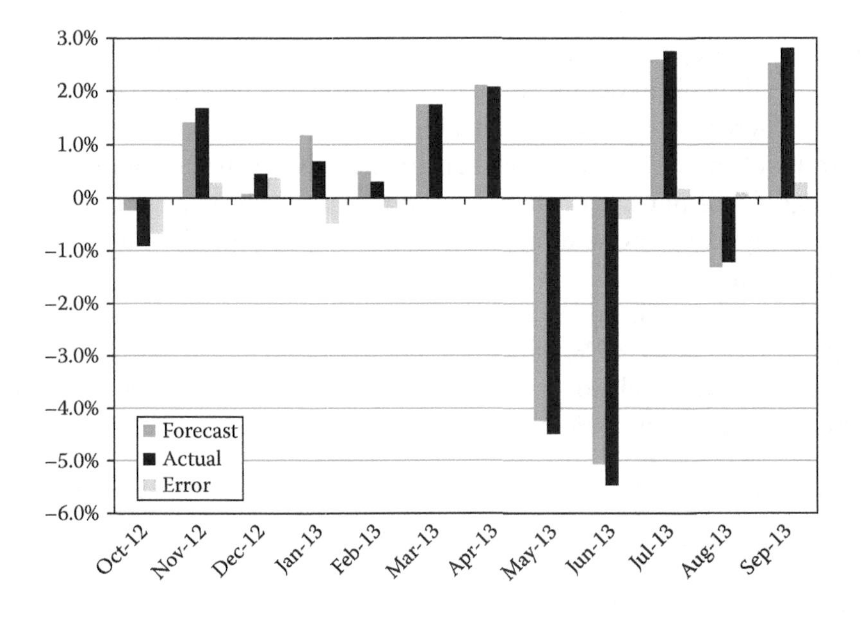

FIGURE 7.5 Forecasted and actual monthly returns of average managers.

the entire period. The errors are quite small compared to the actual returns. Indeed, the annualized tracking error is just 1.16%. In addition, the mean error on a monthly basis is just −0.07%, which is statistically insignificant. Moreover, the errors in May and June of 2013 are equally small. We also note that the predicted cumulative return for the year is 91 basis points and the actual return is four basis points. Therefore, on average, the six risk parity managers underperformed the average style benchmark only slightly, by 87 basis points. Thus, we conclude that the average model portfolio matched the average manager performance quite well.

7.2.4 Individual Risk Parity Managers

We now look at the efficacy of the style analysis to predict individual manager returns over the same out-of-sample period. From a statistical perspective, it is expected that the style models for individual managers cannot match the average model for the "average" manager. Regardless, we evaluate the fit of the style analysis for each manager by measuring the monthly residual between actual manager performance and predicted manger performance.

The first consideration is the accuracy of the forecasts. Table 7.5 shows the mean and standard deviation of the monthly errors. The means range from −0.3% to 0.3% and the standard deviations range from 0.6% to 0.8%. These error terms, while not as good as the average forecast for the average manager, are still excellent. For example, the standard deviation of errors is much smaller than the standard deviation of the actual returns. Also, the t-stat of the mean does not indicate a significant bias in the forecasts.

The second and alternative consideration is the relative performance of managers against their style benchmarks. We emphasize that these style benchmarks are not actual benchmarks used by risk parity managers if they ever used any benchmark. There are many reasons that risk parity managers do not have a widely agreed-upon reference benchmark. First, it is difficult to design risk parity benchmarks because risk parity portfolios are subject to different interpretations. In addition, they could have time-varying notional weights, a trait not shared by traditional benchmarks.* Nevertheless, our style analysis has captured a significant portion of the return variation in all managers. As a result, it is a reasonable analytical exercise to use a style-derived benchmark as a proxy to measure manager performance.

Table 7.6 shows the value added and tracking error of the six managers. The tracking errors range from 2.1% to 2.9%, which are not very different from in-sample tracking errors. However, the value added, or loosely termed alphas show greater variation, ranging from −4.0% (manager A) to 3.6% (manager C). As a consequence, the information ratios

TABLE 7.5 Mean and Standard Deviation of Monthly Forecasting Error and the Related T-Statistics

	A	B	C	D	E	F
Average	−0.3%	−0.2%	0.3%	0.2%	−0.3%	−0.2%
Standard deviation	0.8%	0.6%	0.6%	0.8%	0.7%	0.8%
t-Stat	−1.39	−0.95	1.51	1.07	−1.30	−0.68

TABLE 7.6 The Value Added and Tracking Error of Actual Performance versus the Style Benchmark from October 2012 to September 2013

	A	B	C	D	E	F
Value added	−4.0%	−2.1%	3.6%	3.0%	−3.2%	−2.1%
Tracking error	2.8%	2.1%	2.2%	2.7%	2.4%	2.9%
Information ratio	−1.44	−0.99	1.60	1.12	−1.35	−0.72

* There is no commonly known benchmark for risk parity. Some use cash plus a predetermined return, while others use various 60/40 portfolios as long-term benchmarks.

TABLE 7.7 Cumulative Predicted and Actual Returns from October 2012 to September 2013

	A (%)	B (%)	C (%)	D (%)	E (%)	F (%)
Prediction	2.23	−1.92	0.21	−0.71	0.81	4.88
Actual	−1.94	−4.02	3.65	2.28	−2.52	2.85

have a wide range as well, with manager A delivering an IR of −1.44 and manager C delivering an IR of 1.6.

Of course, it is hard to know precisely the underlying causes for these relative performances. Some of the plausible reasons are (1) style benchmarks are not precise; (2) tactical shifts by managers; (3) deviation from traditional indices in underlying asset class exposures; and (4) volatility timing by managers.

In spite of these drivers of prediction error, the model performed reasonably well. Table 7.7 shows the two sets of return numbers, one for the predicted returns and the other for the actual returns. First, we note the correlation between the two is positive 0.36, indicating the overall ranking of managers is preserved.

Second, consistent with our risk analysis shown in Figure 7.4 and realized performance of asset classes shown in Table 7.4, the model predicts that manager F would have had the best performance (4.88%) and manager B would have had the worst performance (−1.92%). This is because the former has a strong bias toward equity risk and the latter has a strong bias toward interest-rate risk. The predicted spread is close to 7%. In reality, both managers trailed their model portfolios by roughly 2% so the actual spread proved to be close to the predicted spread of 7%. If one had thought that all risk parity managers are a homogeneous group, this kind of return dispersion would dispel that notion.

Third, among the other four managers, our risk analysis (Figure 7.4) also shows manager A has slightly more equity risk exposure than the other three. This is validated by manager A's predicted return of 2.23% shown in Table 7.7. However, manager A underperformed the model by more than 4%.

Finally, managers C, D, and E, are very similar in terms of their risk allocation. Indeed, their model returns are also very similar; however, manager C and D had positive "alpha" while manager E had negative "alpha."

7.2.5 Drawdown Prediction

The last case we shall consider is the prediction of style analysis for May and June of 2013—a period when risk parity strategies in general suffered large drawdowns. Drawdown scenarios are not only of practical importance in risk management, but they are also critical to the theoretical validation of risk models.

Table 7.8 shows the predicted returns based on the style weights and the actual returns from the managers. The prediction is extremely accurate. For example, the prediction for the worst performer (manager B) is −12.0% while the actual performance is −12.5%. Similar precision is true for manager D and F, the two best performers. In hindsight, it is rather apparent why manager B performed worse than others during this period; manager B's portfolio had a significantly higher risk allocation to interest-rate risk and moreover it had the highest notional exposure to inflation-linked bonds, which was the

TABLE 7.8 Predicted and Actual Returns for the Month of May and June 2013

	A (%)	B (%)	C (%)	D (%)	E (%)	F (%)
Prediction	−9.8	−12.0	−8.4	−7.1	−9.6	−7.6
Actual	−11.5	−12.5	−9.4	−6.3	−11.3	−7.0
Difference	−1.6	−0.5	−1.0	0.7	−1.7	0.6

worst-performing asset class. On the other hand, manager D and F benefited, at least on a relative basis, from their higher risk allocations to commodities and equities, respectively. Both commodities and equities, on a risk-adjusted basis, did better than interest rates in May and June of 2013.

The correlation between the predicted and the actual returns was 0.94. Figure 7.6 provides a graphical illustration of the results. The two sets of numbers are close to fitting on a straight line. The R-squared of the fit is 88%. The rank of predicted returns is perfectly preserved in the actual returns. From any perspective, the prediction during this drawdown period appears to be a remarkable success, lending strong support to the validity of the style analysis.

7.2.6 Conclusion

Risk parity multi-asset portfolios should have a balanced, not necessarily equal, risk allocation to three primary sources of risks: equity, interest rate, and inflation. In a previous research note, we concluded that, based on return-based style analysis, some managers are not truly at risk parity due to their concentration bias toward one specific risk.

Reviewing one year of live performance since our original study, albeit a short window, has largely confirmed our results. First, the average model portfolio predicts the "average" risk parity manager with remarkable accuracy. One wonders if this model portfolio

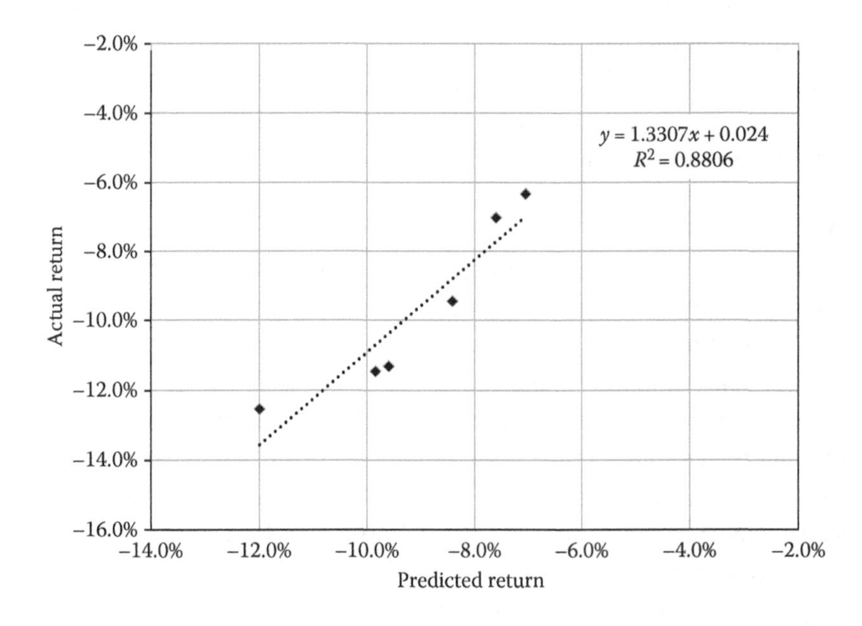

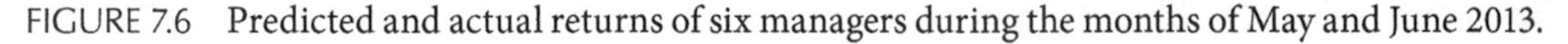
FIGURE 7.6 Predicted and actual returns of six managers during the months of May and June 2013.

could potentially serve as a common benchmark for all risk parity managers. Second, with individual managers, while the tracking errors are larger, the results are generally satisfactory. For example, the model performance and the actual performance for the entire period have significant positive correlation. Moreover, the model correctly identifies the worst- and best-performing managers based on the style weights and market returns of underlying indices.

Finally, for the months of May and June of 2013, the style analysis was remarkably accurate in predicting managers' drawdowns. The risk allocation of managers becomes the dominant factor in determining their absolute and relative returns in this period. For example, manager B had risk concentration in interest rate risk and hence it suffered the most severe losses while manager F had risk concentration in equity risk and suffered the least losses.

In summary, these results suggest that return-based style analysis, combined with a proper risk-based allocation framework, provides valuable insights to investors in determining the true investment style of risk parity managers. In addition, it may also serve to identify common as well as individual benchmarks for risk parity managers.

7.3 "VALUE" OF STOP-LOSS INVESTMENT POLICIES*

A stop-loss investment policy is designed to cut portfolio exposures when an investment strategy has suffered losses exceeding a predetermined threshold. At the extreme, one could close all positions. Normally one would reduce positions proportionally to reduce risk that is, deleveraging. This is only half of the process. A stop-loss policy also needs to determine when to get back in. Often the reentry point is when the strategy stops losing value or when it experiences a small gain.

A stop-loss policy has tremendous psychological value—it gives investors some peace of mind that their investments are protected against significant declines. But could this be the main reason that many retail and institutional investors use stop-loss triggers? Does it necessarily have any investment value?

Whether or not a stop-loss policy actually has investment value is an entirely separate question. The answer, of course, depends on the type of strategy as well as the market environment. In this research note, I analyze both the psychological and investment value of stop-loss policies. In my view, the psychological value is related to the behavioral bias of investors whose utility function appears very sensitive to small losses but strangely less sensitive to large cumulative losses. Thus, an automatic stop-loss rule can overcome that bias, forcing investors to act before significant losses accumulate.

However, this psychological value will lead to positive investment value only if investment returns have certain characteristics. The required statistical properties include positive autocorrelation (losses leading to further losses), negative skewness, and high kurtosis (fat left tail risk). When these conditions are not met, stop-loss policies tend to have minimal investment value such that it would not increase investment returns over time.

---

* Originally written by the author in July 2013.

Guided by these insights, I then apply a simple stop-loss rule to three asset classes: commodities, stocks, bonds, and a naïve risk parity portfolio based on these three asset classes. It is found that the stop-loss rule has investment value for commodities but little or negative investment value for stocks, bonds, and risk parity overall. Its value is akin to that of an insurance policy that is only evident during periods of market stress.

7.3.1 Overcome Behavioral Biases

There are two behavioral biases that are relevant to a stop-loss policy. One is the propensity of selling winners and holding on to losers and the other is an increasing aversion to small losses but less so for large losses. According to Prospect Theory by Kahneman and Tversky, this could be due to the fact that investors behave as if they have a utility function that treats gains and losses differently, rather a logarithmic utility function underlying standard economic theory (Barberis, 2013).

We plot a logarithmic utility function in Figure 7.7. The horizontal axis measures total wealth and the vertical axis is the utility (or satisfaction) that the wealth brings. Focusing on the left side of Figure 7.7, we see that the utility drops sharply as wealth declines.

In contrast, the prospect theory posits that investors have a utility function like the one shown in Figure 7.8. Its shape changes at the reference point, which could be the entry point of an investment such as the price paid on a stock. Instead of thinking about their overall well-being, investors react and feel, rather differently, based on gains and losses relative to the reference point.

Again focusing on the left side of Figure 7.8, we note that the value of utility decreases sharply for small losses. But as the losses become larger, the rate of decline slows down. It seems that investors have extreme aversion to small losses but as losses accumulate, they somehow develop a degree of numbness to losses.

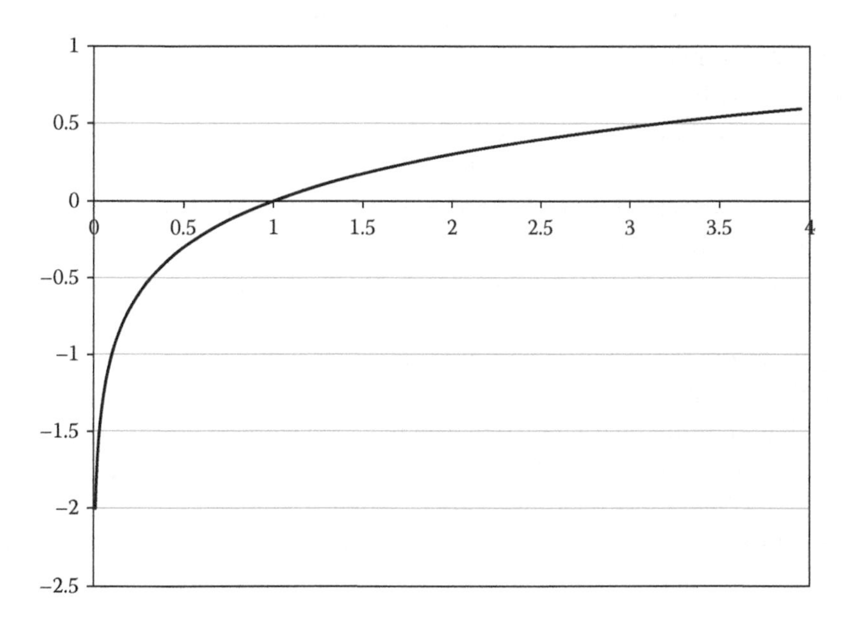

FIGURE 7.7 Logarithmic utility function.

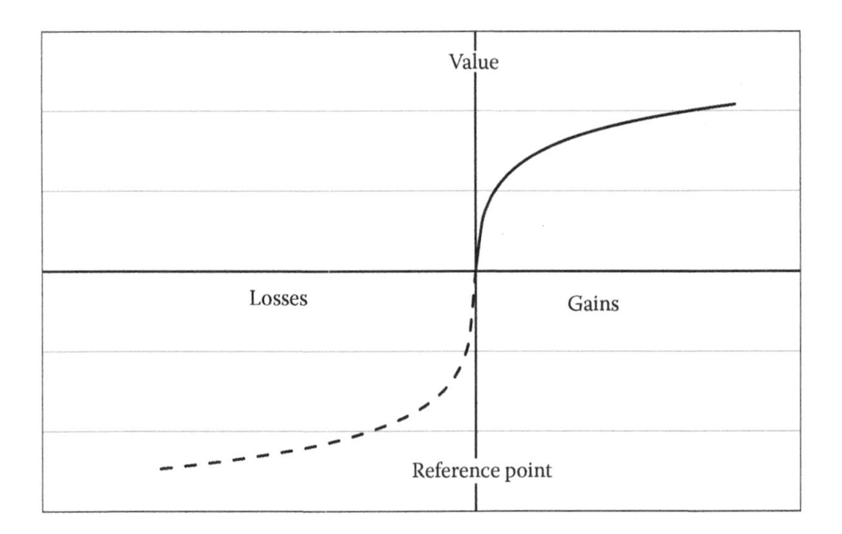

FIGURE 7.8 Utility function of prospect theory.

It is in this context that a stop-loss policy could prevent such predicaments by cutting losses early against investor's biases to keep losers and then becoming emotionally detached to large losses.

7.3.2 When Does Stop-Loss Work?

Of course, this psychological value is just psychological if a stop-loss policy does not have investment value, or worse it destroys investment value. For example, a stop-loss policy could quickly turn into a stop-gain policy if the portfolio limits upside participation during a recovery after the stop-loss policy has been implemented. So the question is when does a stop-loss rule actually stop losses?

Simple reasoning would imply that a stop-loss rule could limit losses if the given investment has the tendency to trend, that is, if losses are followed by further losses. In statistical terms, the return stream has positive serial autocorrelation. On the contrary, if the given investment has the tendency to revert such that losses are followed by gains, stop-loss would detract value. But even if the given investment has no tendency to either trend or revert, stop-loss could still be inferior to buy-and-hold after considering transaction costs.

Another return trait that might serve a stop-loss policy well is the fat tail risk—there are large negative returns occurring more frequently than what a normal distribution implies. If that is the case, a stop-loss policy might be more desirable if the return stream tends to be trending.

7.3.3 Characteristics of Asset Class Returns

Stop-loss can be applied to any investment strategy. In this research note, we are interested in its application to asset class returns and risk parity portfolios. Table 7.9 shows the return characteristics of three major asset classes and a naïve risk parity portfolio.* The monthly

* The naïve risk parity is constructed through equal risk contribution to three asset classes, using only historical volatility but no correlation in determining risk contribution. The total leverage is fixed at 200%.

TABLE 7.9 Return Characteristics of Commodities, Stocks, Bonds, and a Naïve Risk Parity Portfolio

	Cash	Commodity	Stock	Bond	Risk Parity
Return	5.36%	8.01%	10.06%	7.75%	12.01%
Risk	0.96%	20.48%	15.65%	5.33%	11.95%
Sharpe		0.13	0.30	0.45	0.56
Skewness		0.07	−0.44	0.45	−0.29
Ex. Kurtosis		2.46	1.91	3.01	1.84
AC(1)		0.15	0.04	0.14	−0.03
AC(2)		0.02	−0.03	−0.09	−0.05

return series from January 1973 to June 2013 use the GSCI as a proxy for commodities, the S&P 500 index as a proxy for stocks, and the Barclays UST index as a proxy for bonds. The cash return is based on the 3-month UST bill.

While all three assets have had positive excess returns versus cash, bonds have had the highest Sharpe ratio (0.45) followed by stocks (0.30) and commodities (0.13). Due to the diversification benefit, risk parity's Sharpe ratio has been the highest at 0.56.

What is important to the efficacy of a stop-loss policy is the serial autocorrelations of the return series, AC(1) and AC(2) shown in Table 7.9. AC(1) is the autocorrelation with a 1-month lag while AC(2) is the autocorrelation with a 2-month lag. As shown, AC(1) is positive for all three individual asset classes and it is rather significant for commodities and bonds. But AC(2) for both stocks and bonds turns negative and it is marginally positive for commodities. In other words, the asset returns show 1-month momentum or trending but reversal over a 2-month horizon.

This pattern is very different for the risk parity portfolio, however. Both AC(1) and AC(2) are negative, indicating reversal at both lags. One might wonder how this could happen. Statistically, the autocorrelation of the risk parity portfolio depends on not only the autocorrelations of individual assets but also the cross correlation of different assets. If cross correlations are negative (stocks down 1 month leading bonds up next month), then the overall portfolio could mean-revert because the risk parity portfolio has equal risk exposure to bonds and stocks.

What do all these statistics mean? They suggest that stop-loss policy based on a 1-month return horizon could potentially work for individual assets (especially for commodities) and it might not work for more diversified portfolios like risk parity.

What about fat tail risks? According to the excess kurtosis, all asset returns, including those of risk parity, have fatter tails than implied by a normal distribution. Stocks and risk parity also exhibit slightly negative skewness. However, in both cases, there is little trending so a stop-loss policy might not be able to alter its return distribution. This suggests that one might have to look for alternative approaches to reduce fat tail risks.

7.3.4 A Simple Stop-Loss Test

In this section, we investigate a simple stop-loss test for the three-asset classes and the risk parity portfolio. We use the test to enhance our understanding of the analysis above and to observe its efficacy over time. The test is not meant to be comprehensive in data-mining the

TABLE 7.10 Stop-Loss Threshold and Reentry Trigger

	Commodity (%)	Stock (%)	Bond (%)	Risk Parity (%)
Stop-loss threshold	5.0	4.0	1.5	3.5
Reentry trigger	1.0	0.80	0.30	0.70

optimal stop-loss rule parameters such as return horizon, threshold value, reentry trigger, or degree of deleveraging.

Since the trending of returns is more evident in AC(1), we specify the stop-loss threshold on a 1-month horizon, listed in Table 7.10. These numbers are roughly 1-month return standard deviations for the four different investments. When the stop-loss is implemented, we reduce the investments by a quarter* and put the proceeds in cash. This is a more realistic implementation than selling the entire investment. The reentry triggers listed in Table 7.10 are positive and they are set to be one-fifth of the stop-loss threshold. In other words, we require the asset classes or the risk parity portfolio to exhibit a small gain before bringing the investments back to their full exposure.

First, the good news—the stop-loss rule worked very well for commodities. Table 7.11 compares the return statistics with the stop-loss rule to the original buy-and-hold returns. The annualized return increased from 8% to 9.14% while the risk declined slightly. As a result, the Sharpe ratio increased from 0.13 to 0.20. Also, the skewness turned more positive. All these indicate that the simple stop-loss rule stopped some significant losses for commodities. Maybe this result is not entirely surprising since one of the major tools of commodity trading advisors (CTAs) is trend following, which is closely related to stop-loss on the downside.

For the remaining three investments—stocks, bonds, and risk parity, the performance of the stop-loss rule was flat at best. In all cases, there is a reduction in risk since the investments are reduced when the stop-loss is implemented, but the returns and other characteristics show no significant changes.

All these results seem consistent with the autocorrelation analysis, with the exception of bonds. Bonds exhibit strong 1-month trending, just like commodities; yet the stop-loss rule did not add value. This could be because bonds also exhibit strong 2-month reversal. Consequently, the stop-loss rule may have stopped gains during the recovery period before the reentry was triggered.

TABLE 7.11 Commodity Return Comparison

	Commodity	Commodity SL
Return	8.01%	9.14%
Risk	20.48%	19.19%
Sharpe	0.13	0.20
Skewness	0.07	0.31
Ex. Kurtosis	2.46	2.27

* For the three asset classes, this means 75% invested and 25% cash. For the risk parity portfolio, this means the leverage is reduced from 200% to 150%.

In the case of the risk parity portfolio, the overall return also declined while the Sharpe ratio stays near constant. This is consistent with the fact that risk parity returns reverse on a monthly basis.

7.3.5 So When Does Stop-Loss Really Work?

The previous results show that with the exception of commodities, the stop-loss rule has virtually no investment value for the entire 40-year period from 1973 to June 2013, that is, the average gain from the stop-loss rule is close to zero or even negative. But as with any investment strategy, during certain market environments, it could work quite well.

To see how the stop-loss rule works over time, Figure 7.9 plots the ratio of cumulative wealth of an investment with the stop-loss rule to cumulative wealth of an investment without it. Graphically, when the ratio goes up, the stop-loss rule is adding value. Conversely, if the ratio goes down, the stop-loss rule is subtracting value. If the ratio is flat, it is mostly because the stop-loss rule is not on.

Several features of Figure 7.9 are worth noting. First, consistent with the return statistics in Tables 7.11 through 7.14, the end ratio is higher than 1 for commodities and close to 1 for the other three investments. Second, only the ratio for commodities goes up over time in piece-wise fashion while the other three ratios either go down over time or are stagnant. Third, and perhaps the most prominent feature, is the sharp rise of the wealth ratios for commodities, stocks, and risk parity during the global financial crisis. It is rather intuitive that since these investments were trending downward over an extended period of time during the crisis, the stop-loss rule reduced portfolio exposures during most of that

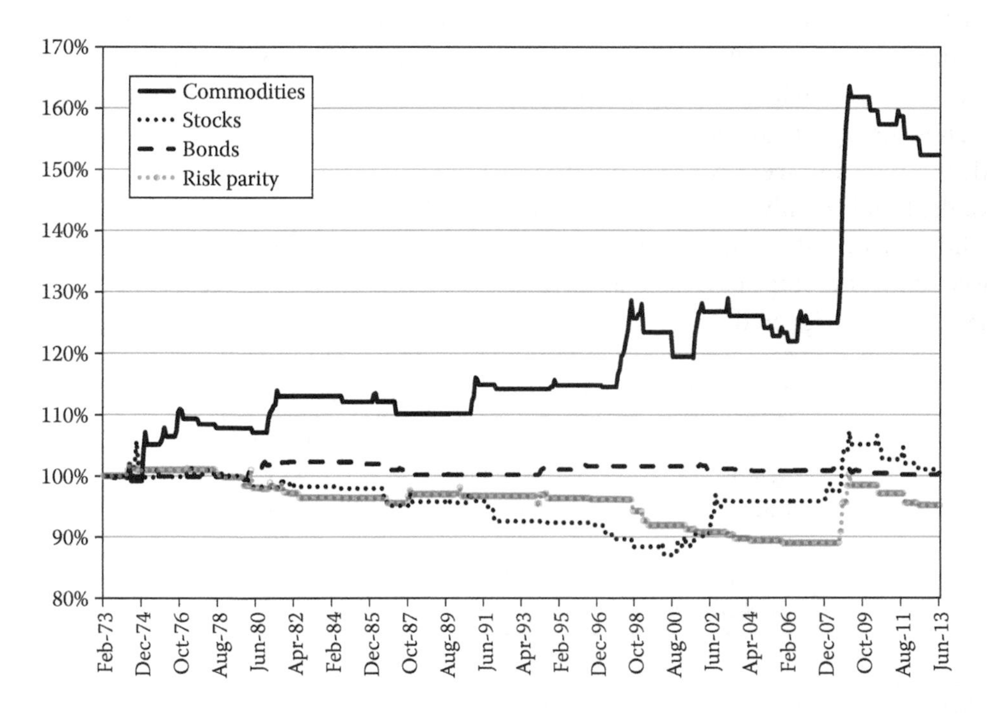

FIGURE 7.9 Ratio of cumulative returns of investments with stop-loss rule to buy-and-hold.

TABLE 7.12 Stock Return Comparison

	Stock	Stock SL
Return	10.06%	10.09%
Risk	15.65%	14.58%
Sharpe	0.30	0.32
Skewness	−0.44	−0.47
Ex. Kurtosis	1.91	1.93

TABLE 7.13 Bond Return Comparison

	Bond	Bond SL
Return	7.75%	7.76%
Risk	5.33%	5.21%
Sharpe	0.45	0.46
Skewness	0.45	0.53
Ex. Kurtosis	3.01	3.20

TABLE 7.14 Risk Parity Return Comparison

	Risk Parity	Risk Parity SL
Return	12.01%	11.87%
Risk	11.95%	11.28%
Sharpe	0.56	0.58
Skewness	−0.29	−0.18
Ex. Kurtosis	1.84	0.86

period, thus generating investment value. It is also noted that this one-time value added was both substantial and short-lived. When the markets recovered in 2009, the stop-loss rule stopped working and it has been losing value for the last few years.* In this regard, the stop-loss rule is akin to an insurance policy, which could lose value over an extended period of time and it pays off only when a crisis occurs. Insurance policies are usually not free. Factoring transaction costs, the stop-loss rule lost value for stocks, bonds, and the risk parity portfolio over the entire period. Another example of insurance is out-of-the-money puts or other long-volatility strategies, which also provide downside protection at a cost equal to that of the option premium.†

Fourth, there are a few other subperiods, during which the stop-loss rule delivered positive results. For example, during the 2001–2003 stock bear market, it added value for stock investments. This episode is to a lesser degree similar to what happened during the global financial crisis. For bond investments, the stop-loss rule added value during 1994 and from 1980 to 1981. Of course, those were the bear market periods for bonds. Again, the stop-loss rule acted like an insurance policy that cushioned the losses for bonds.

* This is consistent with anecdotal evidence that CTAs has not performed well after the global financial crisis.

† It is quite easy to see the similarity of a stop-loss rule and portfolio insurance strategy that is linked to the 1987 stock market crash.

Finally, we note that for the risk parity portfolio, the stop-loss rule seems to have worked only during the global financial crisis. This is because, due to its diversifying nature, the risk parity portfolio rarely suffers losses over extended periods of time. In addition, it exhibits mean-reverting tendencies over the long run, which generally makes stop-loss rules countproductive. Therefore, we do not advocate using a stop-loss policy for risk parity portfolios.

7.3.6 Conclusion

This investment insight analyzes the psychological and investment value of a stop-loss policy. In the context of behavioral biases, a stop-loss policy can be useful in overcoming the propensity of keeping losers and the tendency of becoming less averse or even risk-seeking as losses mount.

However, the investment value of a stop-loss policy is not necessarily positive. We show that a stop-loss policy is expected to add value over time only if the return stream has the tendency to trend or exhibit positive serial autocorrelation. This is true for commodities but not so for other asset classes or a risk parity portfolio. When the returns are on average mean-reverting or just random, a stop-loss policy will lose value over time. In these cases, what is good for your peace of mind is bad for your investments.

However, during financial crises, many investments have the strong tendency to trend downward and therefore a stop-loss policy could have added significant value during a short time period. Hence, one can view a stop-loss policy as an insurance policy, which normally comes with a price. In this case, the price is the negative value added associated with possibly lower return and increased transaction costs.

So how do we explain the popularity of stop-loss when it has a negative investment value? The explanation may partly come from prospect theory which posits that people tend to exaggerate low probability events when they buy insurance or when they gamble. After the global financial crisis, it is natural that this bias of overweighting the probability of a tail risk event has become even stronger. Similarly, after each natural disaster, demand for property insurance increases and premiums often increase. While it is hard to say whether these behaviors are rational, it is important to make investors aware of the potential benefits, pitfalls, and associated behavioral biases related to stop-loss policies.

CHAPTER 8

Risk Parity Everywhere

Much of a Good Thing is a Great Thing

SO FAR, WE HAVE been discussing topics and questions related to risk parity asset allocation portfolios. Readers can be forgiven for thinking that is all there is for risk parity. In fact, this is definitely not the case. The reason for the extensive coverage of asset allocation is that is where most of the fundamental questions arise. In addition, it is probably where the application of risk parity gets a lot of bang for the buck.

Risk parity, or risk budgeting in general, is not simply an investment strategy in asset allocation, even though the original seed of the risk parity approach was sowed in asset allocation for harvesting various risk premiums with a truly diversified portfolio. Risk parity can be beneficial to any investment area where one seeks to combine different return sources together with a balanced risk allocation. Aside from top-down asset allocation, the application of risk parity extends at a minimum to these investment areas:

- Equity portfolios
- Fixed-income portfolios
- Commodity portfolios
- Multifactor risk premiums
- Multifactor quantitative alpha models
- Multi-strategy hedge funds

In all these areas, traditional investment approaches, based on either capitalization-weight indices or ad hoc weighting schemes, often result in portfolios with unintended risk concentration without true diversification. In a majority of the cases, a risk parity approach, based on years of research conducted by my team, leads to better risk-adjusted performance than the corresponding capitalization-weighted indices.

For instance, conventional capitalization-weighted equity indices have risk concentration in country, sector, and individual stocks, causing high risk and low returns in these indices. On the country dimension, the United States currently dominates world equity index with more than 50% of the weight. Decades ago, that role belonged to Japan, which went through its real estate and equity bubbles in the 1980s and then dragged the index down when the bubble burst. Figure 8.1 shows the risk contribution from different countries in the MSCI World Index. As expected, it is dominated by the United States, followed by Japan and the United Kingdom. Many countries have minuscule risk allocations—they simply do not matter.

On the sector dimension, the capitalization-weighted indices are dominated by cyclical sectors whose returns are highly sensitive to business cycles. In contrast, the defensive sectors that provide downside protection to equity investments during periods of weak economic growth or recurring financial crisis have little weight in the indices. Finally, there is often concentration in individual stocks. Most of the capitalization-weighted equity indices are dominated by the top 20% of the names in the indices.

Figure 8.2 displays the risk allocation from the 10 sectors in the S&P 500 index. The defensive sectors, consumer staples, health care, telecom, and utilities account for only about 20% of the risk of the index while cyclical sectors contribute 80% of the risk.

Similarly, conventional fixed-income indices expose investors to risk concentration across country, sector, and quality rating. Conventional commodity indices expose investors to risk concentration in the energy sector as well as individual commodities. Figure 8.3 shows the risk contribution from different countries in the WGBI index. It is dominated by the United States, Japan, and several European countries. Japan actually had and still has the largest weight in the index. However, the volatility of JGBs is much lower than that of USTs, which is the second-largest weight in the index. As a result, the United States has the largest risk contribution and Japan has the second-largest risk contribution.

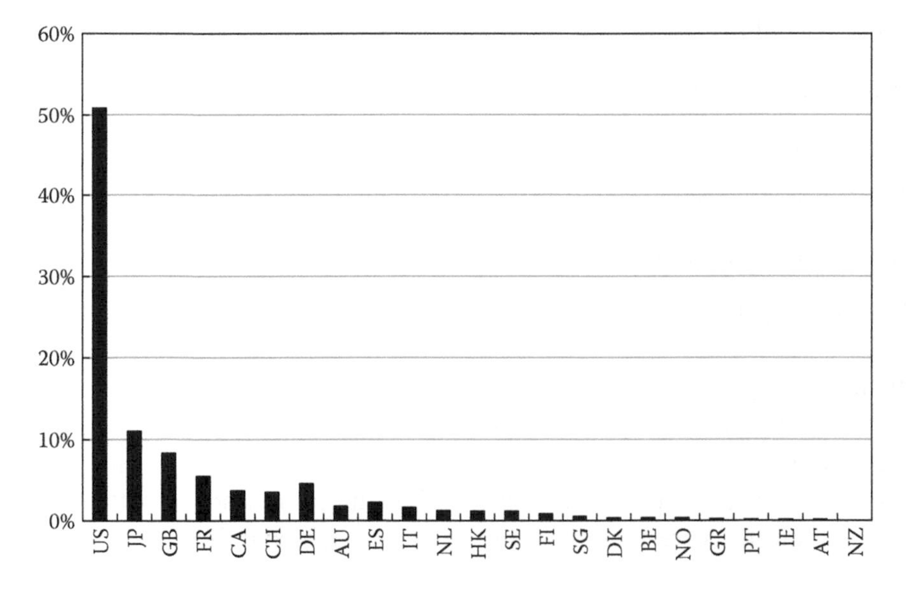

FIGURE 8.1 Country risk allocation to the MSCI World Index.

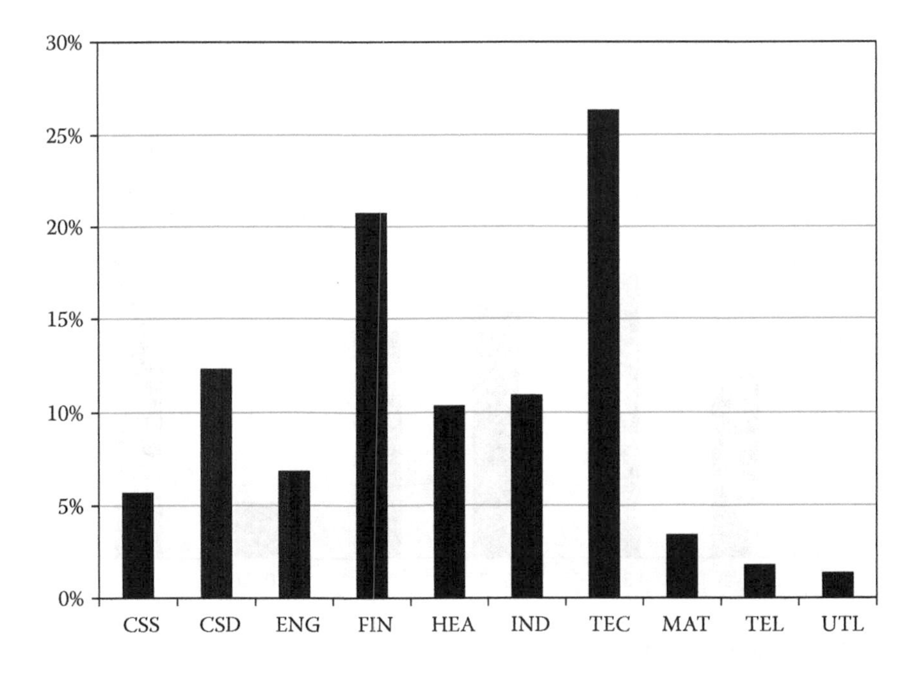

FIGURE 8.2 Risk allocation from the 10 S&P 500 index sectors.

Within individual countries, traditional bond indices might also be undiversified across maturities due to specific government funding needs or irregularities. Risk concentration in particular maturities could expose a portfolio to specific changes of the yield curve. Figure 8.4 displays the risk allocation to the Barclays UST index across different maturities. The weights of maturity buckets are based on the index as of May 2015. The risk allocation to the 10–20 year sector is almost zero while the risk allocation to the 20-plus year sector

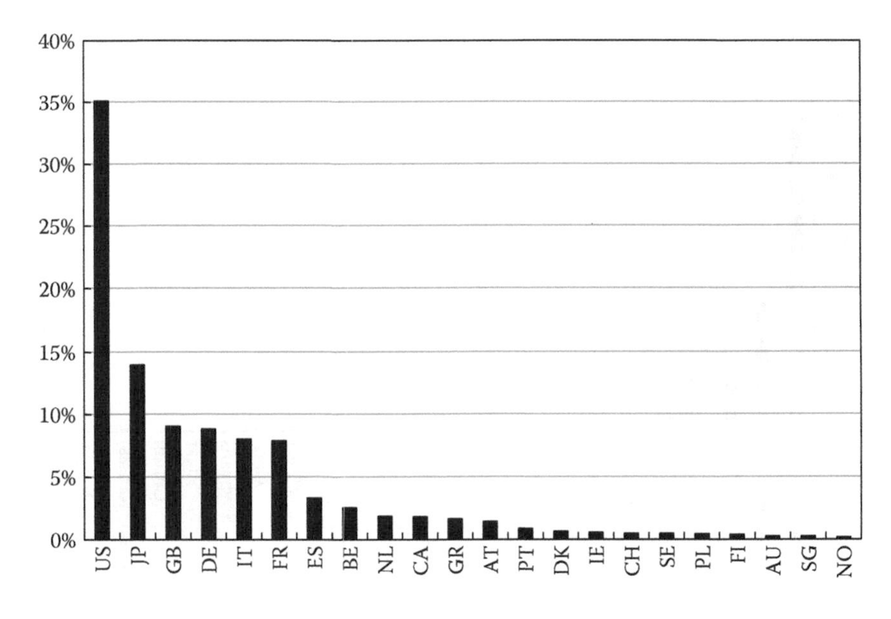

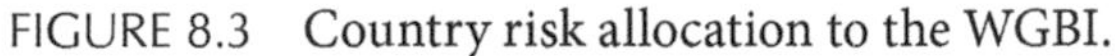
FIGURE 8.3 Country risk allocation to the WGBI.

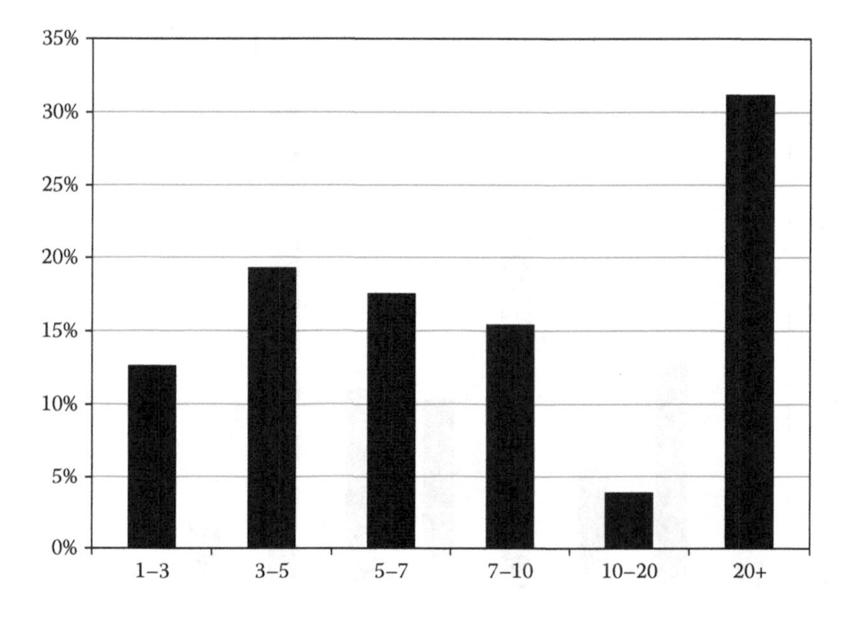

FIGURE 8.4 Risk allocation to the Barclays UST index across different maturity buckets.

dominates. If the long end of the yield curve were to steepen substantially, the index could suffer much high losses than a portfolio that is diversified across all maturities.

The risk allocation of commodity indices is even more concentrated. Figure 8.5 shows the risk allocation of individual commodities to the GSCI index. The only noticeable contributions are all from the energy sector. In fact, the risk contribution of energy commodities accounts for over 95% of total risk! There is little diversification in the index.

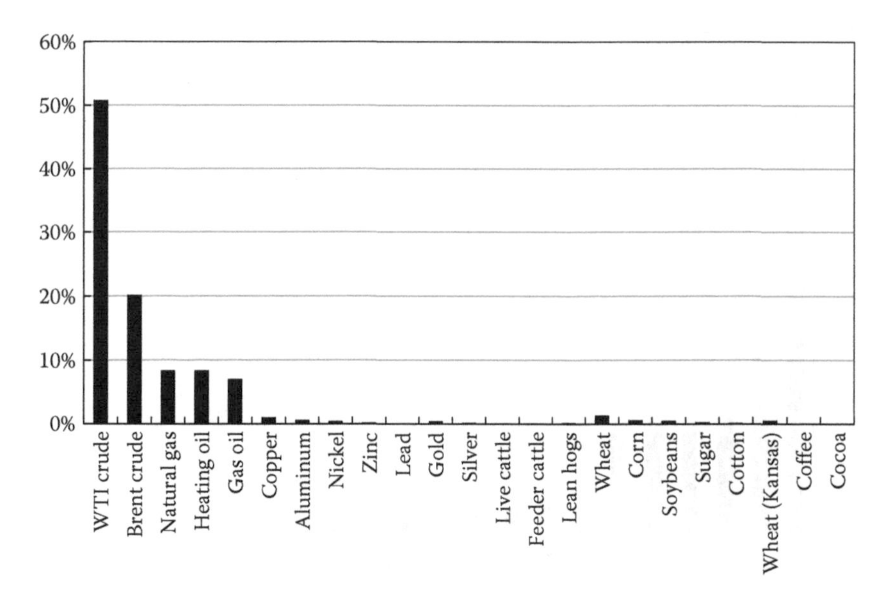

FIGURE 8.5 Risk allocation of individual commodities to the GSCI commodity index.

Many investors have been led to view capitalization-weighted indices or production-weighted commodity indices as "passive" investments. This designation is only true from a narrow mechanical perspective. Yes, there is little trading, in fact little portfolio management required for index investing since capitalization-weighted indices are mostly buy-and-hold portfolios.* However, from an investment perspective, these indices are saddled with strong biases toward certain segments of portfolios such that they are neither diversified nor passive. Unless there are reasons to think these biases lead to better performance than the rest of the portfolios, one should not use capitalization-weighted indices as the vehicles to capture risk premiums.

In this chapter, we present an investment insight on applying a risk parity approach to equity portfolios. Risk parity applied to fixed income and commodity portfolios would similarly result in better risk-adjusted returns. For editorial clarity, we do not provide those insights here in this book. These risk parity portfolios can be used as stand-alone vehicles in equity, fixed income, and commodity. When they are used in a risk parity multi-asset portfolio, the result is a portfolio with risk parity everywhere: both top-down and bottom-up. It is the mother of all risk parity portfolios!

In another insight, we apply the risk parity approach to constructing portfolios with the objective of reaching for yields, which has become an obsession of some investors in today's low interest rate environment. However, reaching for yield by taking more and more credit risk and equity risk could be hazardous to a portfolio's diversification. We outline an alternative approach that balances equity and interest rate risks while reaching for yield.

Another frequently asked question regarding risk parity asset allocation portfolios is, where does it belong in an institutional investor's policy portfolio? Should it be in global asset allocation (GAA) or alternative investments? The answer is it depends. From the perspective of investments, risk parity portfolios are asset allocation portfolios. However, from the perspective of portfolio construction and return characteristics, it actually behaves like a global macro (GM) hedge fund strategy that delivers high risk-adjusted returns. We discuss some features of risk parity portfolios that make it feel like a GM hedge fund and more importantly its advantages over an average GM hedge fund.

The last investment insight in this chapter discusses the application of risk parity portfolios for public and corporate pension plans. We argue that for public pension plans, risk parity should be applied to the total plan level. In other words, 100% risk parity. For corporate pension plans that are managed in an asset-liability framework, the investment policy should consist of two separate parts, one is a liability-matching portfolio and the other is a return-seeking surplus portfolio. Risk parity should be applied to 100% of the surplus portfolio. Realistically, very few plans would embrace risk parity wholeheartedly, even though many plan sponsors would admit privately that it is probably the most sensible thing to do. Oh well, one can hope.

* They thus lack rebalancing alpha generated by regular portfolio rebalancing.

8.1 "GO THE DISTANCE": A MORE GRANULAR APPLICATION OF RISK PARITY*

In the baseball movie *Field of Dreams*, Kevin Costner's character (Ray Kinsella) was inspired by a voice that whispered to him "If you build it he will come." Later in the movie, this same voice also inspired him to "Go the distance" (*Field of Dreams*, 1989).

Today, risk parity products come in a variety of shapes, sizes, and flavors. Despite their differences, any strategy categorized under the risk parity umbrella should hold one core belief in common. This belief is that over the long run investors are not adequately compensated for holding risk concentration in their investment portfolios. Risk parity is a portfolio construction technique designed to eradicate risk concentration by diversifying a portfolio's risk budget. While risk parity managers uniformly agree that risk concentration leads to inefficient portfolios, they differ on which risk concentrations they choose to identify and which risk concentrations they select to address. Indeed, many managers stop short at addressing all risk concentrations.

We think the right approach is to "go the distance" by deploying a deeper and more thoughtful application of risk parity designed to identify and address risk concentrations across multiple dimensions rather than just the asset class dimension.

In this note, we examine the commonly overlooked risk concentrations that exist within the equity asset class. We first identify the dimensions within the equity risk premium where risk concentration is the most injurious. We then show why it is important to "go the distance" to address these risk concentrations by applying risk parity across these risk dimensions with an equity portfolio.

8.1.1 Concentration across the Country Dimension

The simplest way to capture global equity risk premium would be to invest in a broad-based global equity benchmark like the MSCI World Index. This index contains over 1600 stocks across more than 20 different countries. Despite containing a broadly diversified set of constituents, the market capitalization-weighting scheme results in a surprisingly concentrated portfolio. One of the most obvious areas of risk concentration embedded in cap-weighted, global equity indices is in countries. Large countries with many publicly traded corporations tend to exert undue influence on the index. Currently, three countries (the US, the UK, and Japan) account for nearly 70% of the index's total risk. Risk concentration across the country dimension unnecessarily exposes a portfolio to unpredictable country-specific shocks. Country-specific risks include economic risks, political risks, demographic risks, or geographic risks. These events could be something as simple as a change in a country's tax code or as dramatic as a natural disaster. Although the examples are numerous, there are two commonalities. First, the shocks are nearly impossible to predict. Second, investors are not amply compensated for holding country risk concentration in their portfolios.

Risk parity portfolios can identify and address the country risk concentration embedded in capitalization-weighted indices. A broad array of country (albeit cap-weighted

* Originally written by the author in July 2012; coauthored with Bryan Belton, Kun Yang, and Nicholas Alonso.

within country) equity index futures or total return swaps allow investment managers to balance exposure across countries. Rather than constructing a combination of futures and swaps together to minimize the tracking error to the cap-weighted index, a risk parity portfolio can weigh the instruments such that the contribution to equity risk across countries is similar. Risk balancing the portfolio's equity exposure across the country dimension improves the portfolio's efficiency by avoiding a risk concentration that is embedded in traditional equity benchmarks.

But is risk parity across country sufficient in reducing risk concentration, given the fact that correlations among different countries have been on the rise?

8.1.2 Concentration across the Sector Dimension

While risk concentration across the country dimension may be the most obvious, it is not the only concentration embedded in cap-weighted equity indices. Actually, it is not even the most important dimension of risk concentration. That honor belongs to risk concentration across the sectors. Currently, three sectors (financials, technology, and industrials) account for 50% of the index's risk budget. In addition to sector-specific risks, the risk concentration across sectors in cap-weighted indices is mainly concentrated across the cyclical versus defensive groups that expose portfolios to global growth risks. Globalization of trade and capital has caused the synchronization of business cycles of many countries. Risk concentration in the cyclical sectors of cap-weighted indices results in large drawdowns when the global economy weakens, which lowers risk-adjusted returns over the long term.

To see this, Table 8.1 compares the rank of each sector's risk contribution to the MSCI World Index with a measure of the sector's cyclical versus defensive score. We measure a sector's cyclical score by regressing each sector's returns against the Leading Economic Indicator series. Sectors with strong positive correlation to the Leading Economic Indicator series are assumed to be cyclical in nature while sectors with lower correlation to the business cycle are assumed to be more defensive. The ranking scale 1–10 identifies sectors from the most cyclical (1) to the least cyclical (10). Table 8.1 indicates that the rank correlation

TABLE 8.1 Rank Scores of Sectors in Risk Allocation to the Index and Cyclicality

Sector	Risk Weight Rank	Cyclical Score Rank
CSD	4	5
CSS	7	10
ENG	5	4
FIN	1	1
HEA	8	9
IND	3	7
MAT	6	6
TEC	2	2
TEL	9	3
UTL	10	8

between sector risk weights in the index and sector cyclical scores is +0.59. This suggests that the sectors with the largest risk weights in the index are generally also the most cyclical in nature.

While holding risk concentration in a portfolio is rarely beneficial, a risk concentration that affects a portfolio's balance across a systematic dimension like cyclical and defensive or risk on and risk off is particularly perilous. While the performance of equity risk premium is inherently linked to the business cycle, equity exposure concentrated in cyclical sectors tends to move the outcomes even more to the tails of the return distribution—an inefficient approach for capturing equity risk premium.

Most risk parity managers seek to build a portfolio that delivers balanced performance around various economic growth outcomes. For many managers, this exercise starts and ends with balancing the risk contribution between equity risk premium and interest rate risk premium. What our research suggests is that these same imbalances around economic growth risks that exist due to a portfolio's stock/bond weighting also exist because of the way sectors are weighted in the portfolio's exposure to equities.

8.1.3 Concentration across the Stock Dimension

In addition to being concentrated across the country and the sector dimension, cap-weighted indices are also concentrated across individual stocks. This risk concentration introduces unnecessary idiosyncratic risk, which reduces the efficiency of the equity risk premium capture.

Despite containing over 1600 securities, the top 15% of the holdings in the MSCI World Index account for over 60% of the index's total risk (Table 8.2). This risk concentration is partly driven by past performance. While momentum could be a short-term alpha factor, it simply makes no sense to allocate a majority of a portfolio's risk budget to a handful of recent winners. Over time, not many large companies can maintain their competitive advantages as competition intensifies and technology changes.

8.1.4 The Most Important Dimension to Address

We identify risk concentrations across countries, sectors, and stocks that are common in traditional equity exposures. In a risk parity global equity portfolio, we thus strive to mitigate these risk concentrations where practical so the risk allocation is balanced in these three dimensions simultaneously. The result is a much more efficient portfolio of capturing

TABLE 8.2 Risk Contribution From the Group of Stocks in the MSCI World Index

	Stock 1–250 (%)	**Stock 251–750 (%)**	**Stock 751–1635 (%)**
Risk contribution	62	27	19

TABLE 8.3 Portfolio Returns and Risks from 1995 to 2012

	MSCI World Index (%)	**Risk Parity Global (%)**
Return	6.56	9.73
Volatility	14.73	12.37

equity risk premium than the capitalization-weighted index. As shown in Table 8.3, risk parity global equity had a higher return and lower risk than the MSCI World Index from 1995 to 2012.

In determining what is "practical" and beneficial, it is also useful to identify the value of addressing each area of risk concentration individually. To conduct this analysis, we have designed a step-wise attribution framework, which separates the benefits of applying risk parity portfolio construction techniques across the country, sector, and stock dimensions. Table 8.4 summarizes the annualized value added of a risk parity global equity portfolio relative to the cap-weighed MSCI World Index.

Slightly more than half of the value added from deploying risk parity to construct equity exposures is coming from balancing the risk contribution across sectors. Balancing risk across the countries has been the second most important dimension (30% of the value added) while balancing it across stocks has been the third most important dimension. These empirical results are consistent with our qualitative prior. The inefficiency of any particular risk concentration is conditioned upon whether the concentration causes the portfolio to unintentionally become directional across systematic risk factors. As we have shown in this note, traditional equity exposures tend to be cyclically directional across both economic growth risks as well as the "risk on and risk off" sentiment. Balancing the contribution to risk across sectors reduces sector risk concentration and consequently reduces those directionalities. While reducing concentration across countries and stocks is beneficial, its impact is lower than that of addressing sector concentrations because these concentrations are less directional across systematic risk factors.

Table 8.5 shows the correlation between the value added from applying the risk parity portfolio construction technique to the sector, country, and stock levels with the monthly index returns for both the MSCI World Index as well as our risk parity global equity portfolio. The value-added return attributable to applying risk parity to the sector dimension has a strong negative correlation with the returns of both the MSCI World Index as well as the risk parity portfolio. This suggests that balancing risk across sectors appears to be "hedging out" some of the systematic directionality inherent in equity risk premium. Conversely, balancing the risk contribution across strictly the country dimension does not appear to offer this same hedging benefit.

Risk parity equity portfolios are not merely a defensive strategy. Risk balance in equity portfolios does not prevent them from participating to the upside in bull market

TABLE 8.4 Risk Parity Valued Added 1995–2012

	Total VA	Sector VA	Country VA	Stock VA
Annualized return	2.69%	1.43%	0.85%	0.41%
Reward/risk	0.47	0.40	0.24	0.27

TABLE 8.5 Correlations of Value Added with Broad Equity Market

	Sector VA	Country VA	Stock VA
Correlation with MSCI	−0.69	−0.06	−0.41
Correlation with RP	−0.51	0.22	−0.22

equity cycles. Rather, a risk-balanced equity portfolio represents a more efficient way to capture equity risk premium. Our experience is that the risk parity investment is more consistent in experiencing shallower drawdowns and quicker recoveries. Our attribution framework indicates that a majority of this benefit is coming from risk balancing the portfolio across the sector dimension. Sector balance provides the greatest benefit because it not only reduces the idiosyncratic risk of sector concentration, but it more importantly reduces the systematic risk of being directional across the growth risk or risk on risk off dimensions.

8.1.5 Conclusion

Eradicating any form of risk concentration through diversification is the most efficient way to achieve long-term wealth creation. Philosophically, most risk concentrations serve as an impediment to achieving consistent performance results. As a result, we seek to identify and address as many areas of risk concentration as practical. To accomplish this, we apply our risk parity portfolio construction technique not only to balance the risk contribution across asset classes, but also to balance the risk contribution within asset classes. In this note, we identify the three most important pockets of risk concentration that exist in traditional equity exposures. While many risk parity managers are actively reducing risk concentration across the country dimension, this approach fails to address the risk concentration that exists at the sector and stock levels. In our view risk parity approach should "goes the distance" by applying the risk parity principle to a deeper, more thoughtful level. In the case of building exposure to equities as well as fixed income and commodities, we balance the risk contribution across multiple risk dimensions, to capture equity, interest rate, and inflation risk premiums more efficiently than other more risk-concentrated approaches.

8.2 REACHING FOR YIELD: THE RISK PARITY WAY*

Like moths flocking to flames, investors at times reach for yield. While this attraction to higher-yielding assets is not always fatal, investors can get burned by it. The reason is quite simple: some either underestimate or are oblivious to the risks associated with reaching for yield. On a microlevel, investors suffer losses or underperform when carry- or yield-reaching strategies unwind. On a macrolevel, reaching for yield could lead to credit excesses and consequently, a financial crisis when credit bubbles burst.

The act of reaching for yield is an active decision to swap low-yielding assets (often government bonds) for HY assets (lower quality credit bonds). With global interest rates trending lower and lower since 2008, reaching for yield has shown no sign of abatement. Before we get too close to the flames, it is probably time to highlight and analyze the investment risks embedded in some common approaches in reaching for yield and to explore ways to properly manage these risks.

In this research note, we show it is important to recognize that bond yields can be decomposed into several components including the risk-free rate, term spreads, and credit

* Originally written by the author in May 2013.

spreads.[†] We show that the main culprit of the current environment's low yields is the zero or near-zero short-term interest rates across many developed countries. Hence, reaching for yield today means either to increase the term spread or credit spread. We demonstrate that if one is to reach for yield, a diversified and robust approach is to balance the two sources. In other words, reach for yield the risk parity way.

8.2.1 Why Bond Yields Are So Low?

Bond yields are low almost everywhere. Figure 8.6 shows the historical yields for 3-month T-bills, the US 10-year treasury bonds, the US corporate bonds, the US HY bonds, and emerging market bonds. Currently, they are all at or near historical all-time lows. But the most noticeable case is the 3-month T-bill rate, which has been near zero since the end of 2008.

In investing, everything is relative. If we measure the term spread of the 10-year bond versus the risk-free rate, the picture is a little different. As shown in Figure 8.7, the yield-curve slope between 3-month and 10-year ranges from near zero to 4%. The current level is near 2%, close to the historical average. So relatively speaking, the term spread is not too low.

For the remaining bond markets, we measure their credit spreads versus the 10-year yield, shown in Figure 8.8. They are quite tight compared to historical levels. Two features are worth noting. First, the recent tightening of spreads is highly synchronized across the three bond markets, indicating a potential common reach for yield theme. Second, in all three cases, the spreads are nevertheless wider than the extreme low levels reached during periods before the global financial crisis.

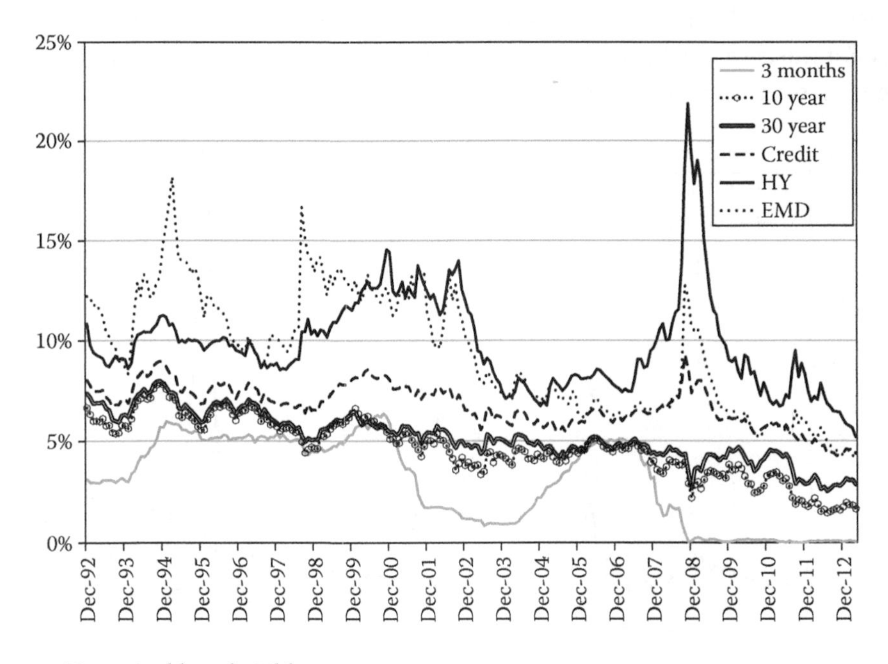

FIGURE 8.6 Historical bond yields.

[†] There are other specific perhaps less-common components such as option-related spread and illiquidity spread. In this research note, we focus only on term spread and credit spread.

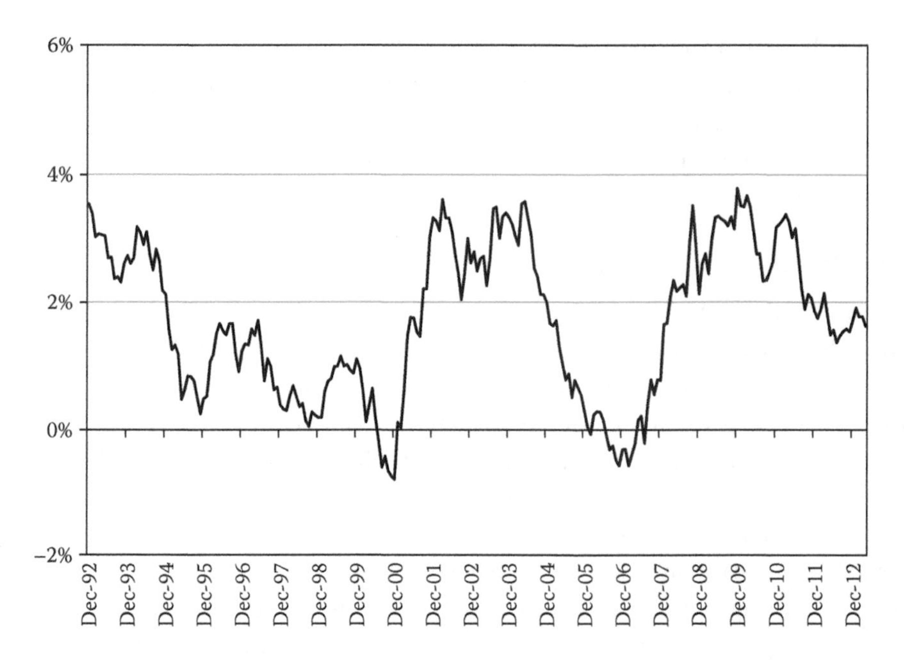

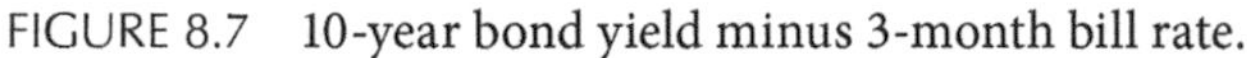
FIGURE 8.7 10-year bond yield minus 3-month bill rate.

While neither term spread nor credit spread is at their historical lows, it is rare that the risk-free rate would get any lower. But if inflation is low or negative, a 0% risk-free rate may not prove to be low enough. A case in point is Japan, which has had a ZIRP and deflation for decades. Ironically, the real interest rate in Japan is actually quite high.

This is not the case outside of Japan. Figure 8.9 plots the 3-month UST bill rate, together with the year-over-year percentage change of the US core CPI, and their difference. The

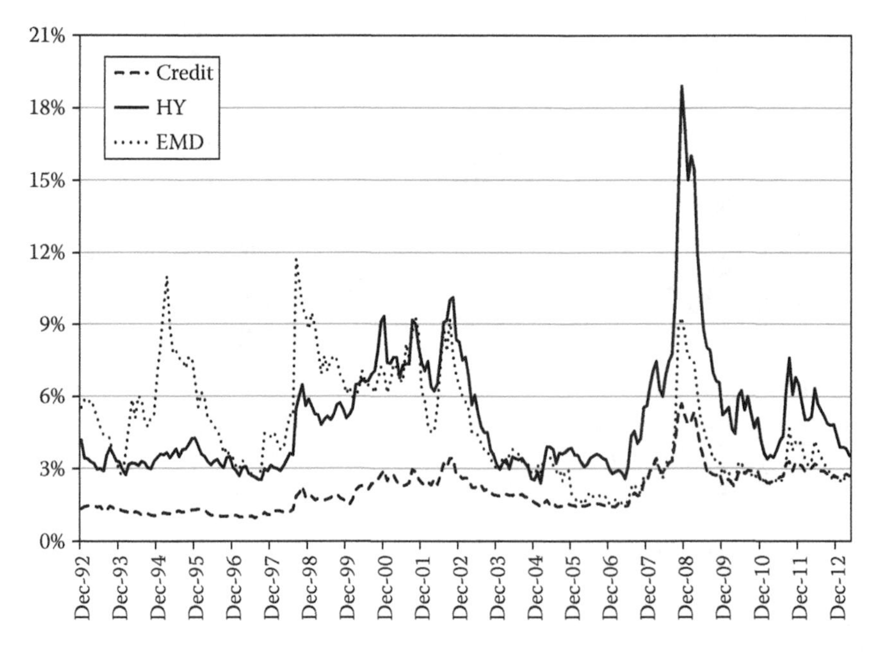

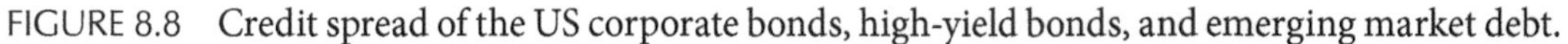
FIGURE 8.8 Credit spread of the US corporate bonds, high-yield bonds, and emerging market debt.

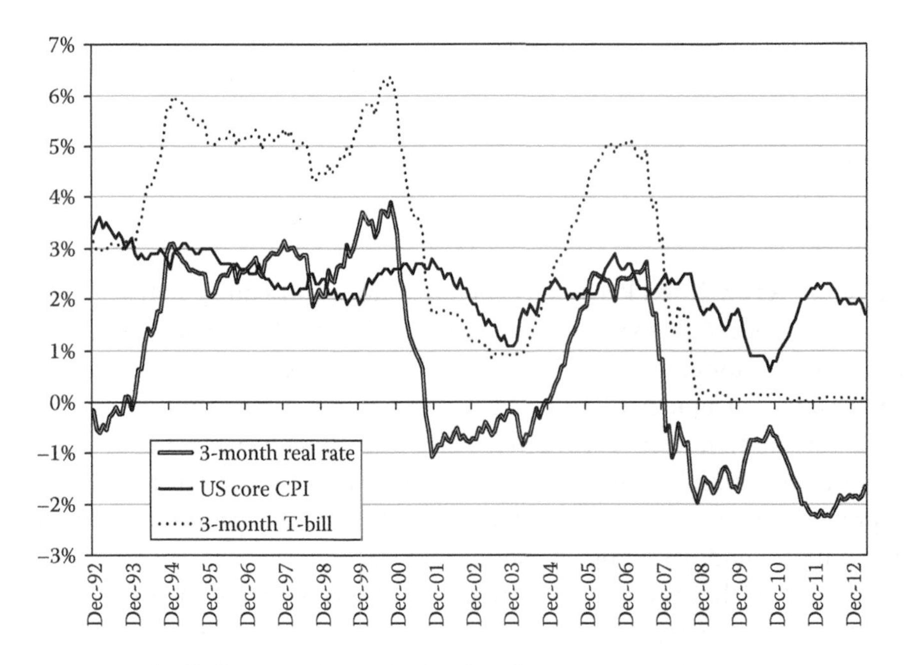

FIGURE 8.9 3-month T-bill rate, US core CPI, and real rate.

real interest rate has been negative since 2008 when the Fed cut the Fed Funds rate to near zero. Having rebounded from under 1% during 2010, the core CPI in the United States is now at 1.7%. As a result, the real short-term interest rate in the United States is close to −1.7%. Figure 8.9 also reveals that negative real interest rates are not unprecedented. It happened around the two recessionary periods in our sample,* one from 2001 to 2003 and the other from 1992 to 1993. During these recessions, the Fed cut short-term interest rates below inflation to spur economic growth, resulting in negative real rates. After the economy recovered from the recession, the Fed normalized policy by raising short-term interest rates above inflation.

So far, this has not been the case this time. As the US economy struggles to recover from the global financial crisis, the Fed has kept the ZIRP for a long time and it has committed to keep it until there are signs of a substantial recovery in the US job market, which could still be some time away.

Therefore, the search for yields is on. Anything offering positive real yields looks better than cash, which has a negative real yield of −1.7%!

8.2.2 The Risk/Return Trade-Off of Reaching for Yield

There are three traditional ways of reaching for yield. The first is to increase term spread by extending portfolio duration; the second is to increase credit spread by lowering credit quality; the third is to invest in high dividend stocks. From a macroeconomic perspective, the latter two approaches are similar as they both assume additional growth risk. If economic growth strengthens both spread sectors and equity markets should do well

* We chose the period from 1992 based on the data availability of EM debts.

TABLE 8.6 Correlations of Term Premium and Credit Premiums

	US 10-year	Corporate Bond	High Yield	EM Debt
US 10-year	1.00	−0.53	−0.54	−0.34
Corporate bond	−0.53	1.00	0.85	0.57
High yield	−0.54	0.85	1.00	0.65
EM debt	−0.34	0.57	0.65	1.00

and if economic growth weakens both spread sectors and equity markets would suffer. The performance of credit spread exposure is highly correlated with equity market performance. From a fundamental perspective, reaching for credit spread or dividend yield exposes investors to heightened growth risk.

On the other hand, the term premium tends to be countercyclical. It often behaves oppositely to credit spread and equity markets. If economic growth is stronger than expected, the term premium tends to be low as bond yields rise and if the economic growth weakens, the term premium tends to be high as bond yields fall.

Table 8.6 displays the correlations between the excess return of the US 10-year treasury (vs. cash) and the excess returns of the three credit sectors (vs. the 10-year treasury). The three credit sectors are highly correlated among each other and they are all negatively correlated with the 10-year treasury.

Therefore, from a diversification perspective, when reaching for yield, we recommend that investors balance term spread with credit spread to balance the portfolio's exposure to growth risk. This proposition should sound familiar to readers with an appreciation of risk parity investing. In a sense, if one wants to reach for yields, one must balance interest rate risk premium with credit premium (a close cousin of equity risk premium).*

8.2.3 How to Increase Term Spread

Before discussing how to combine term spread and credit spread together, we address the various ways of increasing term spread. Investors have multiple choices when increasing credit spread: by increasing allocation to corporate bonds, HY bonds, or emerging market debt, at the expense of lower-yielding assets. On the other hand, there seems to be only one way to increase term spread. That is by shifting into longer-duration bonds.

While duration extension is a common way to reach for yield, it is not the only way nor is it the most efficient way. From a risk-adjusted return perspective, longer-duration bonds are often not as attractive as shorter-duration bonds. For example, 30-year UST bonds have had a lower Sharpe ratio than 10-year treasury bonds. Table 8.7 shows the return and risk statistics of the two benchmark bonds from January 1993 to April 2013. While the risk almost doubles by extending the maturity from 10 years to 30 years, the return only increases by 20%. As a result, the Sharpe ratio of the 30-year bond is 0.48 while that of the 10-year bond is 0.63.† This phenomenon of lower duration bonds delivering better risk-adjusted performance is analogous to the low volatility phenomenon in equity market

* One should also consider the risk of inflation in reaching for yields by comparing real yields of inflation-linked bonds. However, for the purpose of this exposition, we focus only on nominal bonds.

† The cash return is about 3% for the period.

TABLE 8.7 Return, Volatility, and Sharpe Ratio

	US 10-year	US 30-year
Return	7.01%	8.62%
Volatility	6.33%	11.59%
Sharpe ratio	0.63	0.48

where low volatility stocks outperform, often in terms of absolute return, high-volatility stocks.

The investment implication of this phenomenon is that there is an alternative approach to longer-duration bonds when reaching for term spread. In fact, one can increase their term spread exposure by simply increasing their allocation to existing bonds by using leverage. For example, if a 10-year bond has a term spread of 2%, then with 2× leverage, the term spread would be 4%. This yield is actually higher than the 30-year bond yield currently at 3.17% and we see from Table 8.7 that the two positions, one with 10-year bonds 2× levered and the other with 30-year bonds, have similar ex post-return volatilities. The combined effect is that 10-year bonds levered 2× offers a higher risk-adjusted yield than 30-year bonds.

One of the motivations for reaching for yield is to invest in assets with positive real yield and thus maintain real purchasing power. With short-term interest rates at zero, one can enhance yield by being long steep parts of the yield curve and at the same time short cash, effectively achieving positive real yield.

To summarize, there are two ways to add term spread to a portfolio. One is to shift to long-duration bonds and the other is to leverage existing bonds. It is often the case that the latter approach yields better results since long-duration bonds are typically less risk efficient than bonds of shorter durations. Regardless of how you choose to get term spread exposure, it is important to balance term premium and credit premium when reaching for yield.

8.2.4 A Risk Parity Fixed-Income Portfolio

To illustrate this risk-balanced approach, we must consider where the starting portfolio is in terms of its yield. Suppose an investor has a portfolio with 100% treasury bonds, then he or she should make shifts into credit sectors with higher yields. But suppose an investor has a portfolio with 100% in high-yield bonds to begin with, he or she probably should reach for quality instead of yield!

In our view, a strategic fixed-income portfolio should balance between term premium and credit premium. We construct one with the four fixed-income assets used in our previous examples. The portfolios weights are given in Table 8.8. These weights give rise to approximately equal risk allocation to the four assets based on their historical volatilities

TABLE 8.8 Portfolio Weights of a Risk Parity Portfolio

	US 10-year (%)	Corporate (%)	High Yield (%)	EM Debt (%)	Portfolio (%)
Weight	35	30	20	15	100
Yield	1.67	2.60	4.95	4.09	2.97

and correlations. Since the 10-year treasury and the IG corporate bonds have lower volatilities, their weights are much higher than the other two assets. The risk-adjusted return of this portfolio has been very attractive since 1993. The annual return is 8.1% while the annualized volatility is 5.7% and the Sharpe ratio is 0.88, which is higher than the individual Sharpe ratios of all four asset classes.

Even though this portfolio has a substantial allocation to credit, its current yield is just below 3%. This is low considering the average yield of this portfolio since 1993 is 6.8%! Figure 8.10 shows the yield decomposition of the risk-free rate, term spread, and credit spread over time. Again, the risk-free rate accounts for the majority of the decline in the total yield over the recent period. The term spread was occasionally negative when the treasury yield was inverted. Presently, the term spread and the credit spread are balanced at 1.6% and 1.3%, respectively.

8.2.5 Reaching for Yield the Risk Parity Way

To enhance the yield of the risk parity portfolio in Table 8.8, a natural impulse would be to eliminate the allocation to UST bonds, which has the lowest yield, and move down the credit quality spectrum. At the extreme, this could be akin to moths flocking to flames, exposing the portfolio to recessionary risks. The risk parity way of reaching for yield is to increase both term spread and credit spread through leverage. We compare both approaches in this section.

Starting with the risk parity portfolio shown in Table 8.9, we construct three portfolios with higher yields, by eliminating treasury bonds and investing all proceeds in one of the other three assets, respectively. Portfolio C in Table 8.9 has a 65% allocation to corporate bonds, a 20% allocation to high-yield bonds, and a 15% allocation to EM

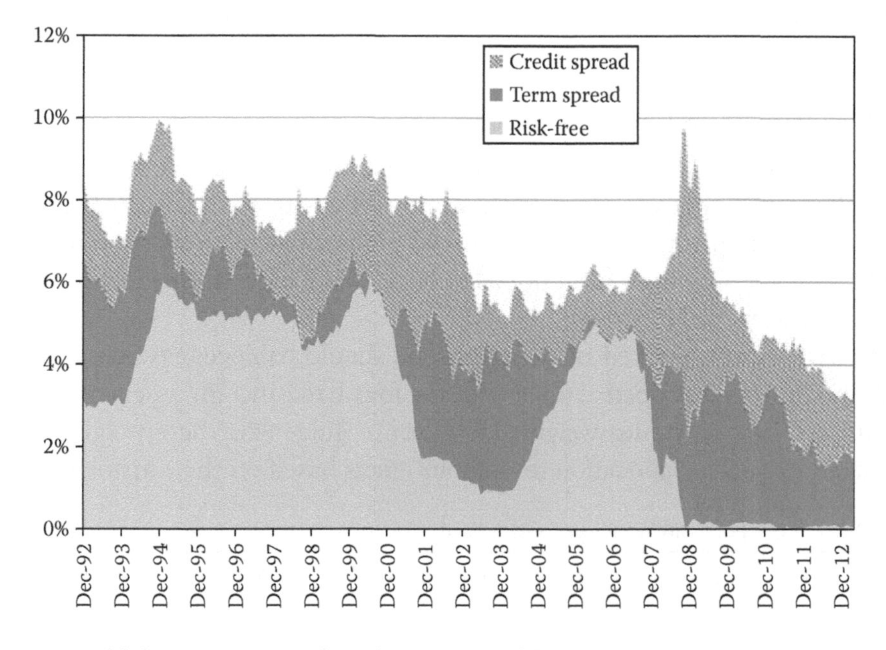

FIGURE 8.10 Yield decomposition of a risk parity portfolio.

TABLE 8.9 Portfolio Weights of Different Yield Enhancement Choices

	US 10-year (%)	Corporate (%)	High Yield (%)	EM Debt (%)	Total Weight (%)
Risk parity	35	30	20	15	100%
Portfolio C	0	65	20	15	100%
Portfolio H	0	30	55	15	100%
Portfolio E	0	30	20	50	100%
Risk parity L	49	42	28	21	140%

debt. Portfolio H has a 30% allocation to corporate bonds, a 55% allocation to high-yield bonds, and a 15% allocation to EM debt. Portfolio E has a 30% allocation to corporate bonds, a 20% allocation to high-yield bonds, and a 50% allocation to EM debt. These three portfolios increase yields with a higher exposure to credit spread while keeping the term spread the same.

Another approach is to increase yield with both higher term spread and credit spread. This is represented by portfolio "Risk Parity L," or leveraged risk parity portfolio, formed by levering the risk parity portfolio by 40%. The portfolio characteristics of this portfolio, in all aspects, appear superior to all other portfolios!

We first look at the current yields and their components of the portfolios in Table 8.10. First, we note portfolio C only increased yield by 30 bps from 3% to 3.3%, since corporate yields are not much higher than treasury yields. Second, the remaining three portfolios all increased yield significantly. Portfolio H (with 55% in HY bonds) and the levered risk parity portfolio both have yields above 4%.

We next look at the historical returns and risks of the portfolios. Figure 8.11 plots the total returns and annualized volatilities of the portfolios since 1993. The risk parity portfolio delivers higher returns with lower volatility compared to Portfolio C. Similarly, the levered risk parity portfolio dominates Portfolio E and H with higher return but lower volatility. As a result, the Sharpe ratios of the risk parity portfolios are higher than the Sharpe ratios of all other portfolios.

Finally, we look at the annual performance of the portfolios, displayed in Figure 8.12. Because risk parity portfolios are balanced between term premium and credit premium, they perform better during periods of weak economic growth and financial stress. This outperformance was particularly strong in 2008. They also performed better during the emerging market crisis of 1998, and during the period from 2000 to 2002. Conversely, the risk parity portfolios tend to lag other portfolios during periods of economic and financial market recovery. This was the case in 2009 following the global financial crisis of 2008, in

TABLE 8.10 Yields and Yield Components of Different Portfolios

	Risk-Free (%)	Term Spread (%)	Credit Spread (%)	Total (%)
Risk parity	0.0	1.6	1.3	3.0
Portfolio C	0.0	1.6	1.6	3.3
Portfolio H	0.0	1.6	2.4	4.1
Portfolio E	0.0	1.6	2.1	3.8
Risk parity L	0.0	2.3	1.8	4.1

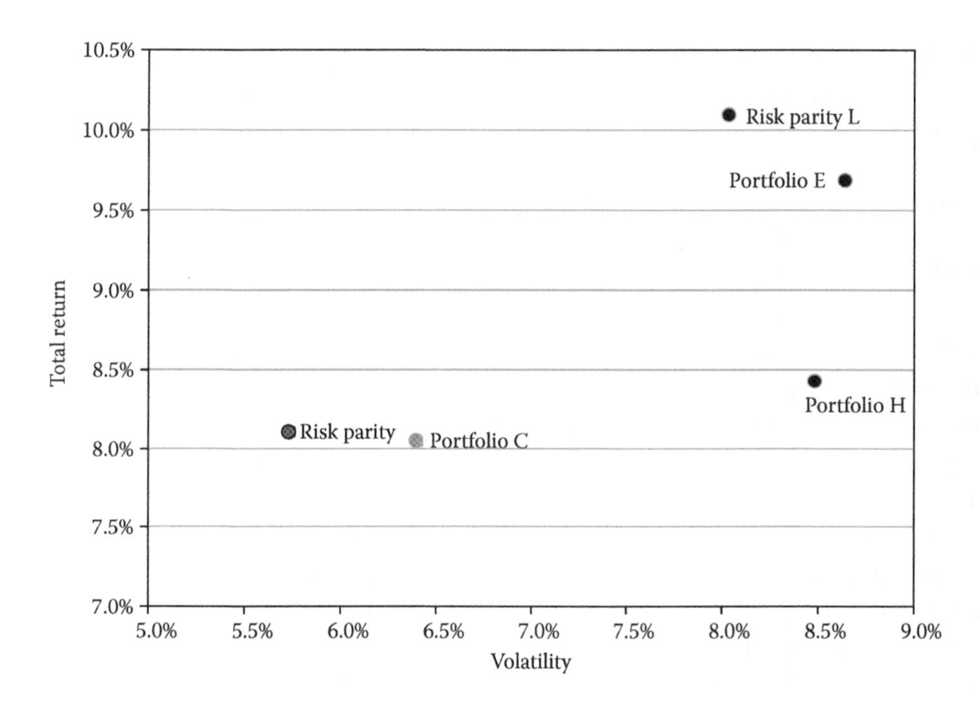

FIGURE 8.11 Risk/return of five portfolios.

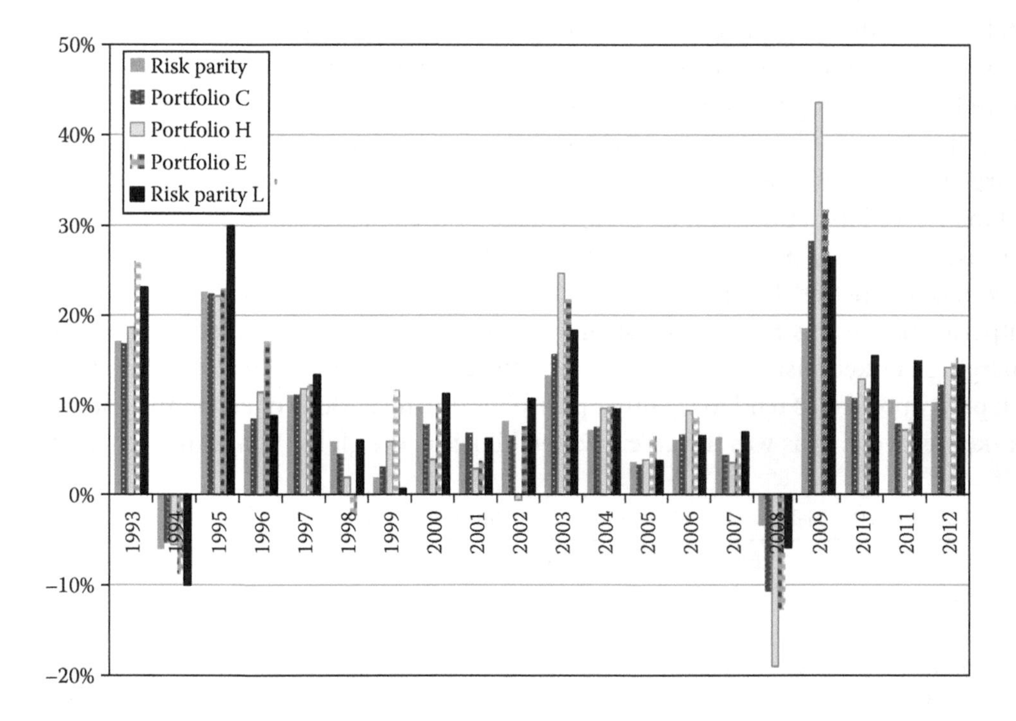

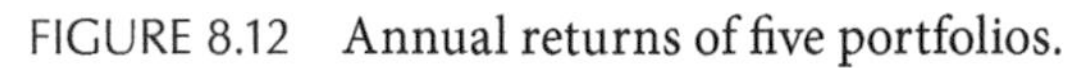
FIGURE 8.12 Annual returns of five portfolios.

1999 after the emerging market crisis, and in 2003 when the market recovered from the tech bubble and a mild recession.

8.2.6 Conclusion

Given the ultralow bond yields in many fixed-income markets, many investors feel the need to reach for yield. With risk-free rates near zero, reaching for yield entails investing in assets with either higher-term spreads or credit spreads, or both. We argue that the risk exposure of credit spread is cyclical, or equity-like while the risk exposure of term spread is countercyclical and a combination of the two using a risk parity approach provides impactful diversification. The traditional method to reach for term spread is to extend the maturity to longer-duration bonds. We show that an alternative and more efficient method is to leverage intermediate duration bonds. The comparison of different yield enhancement strategies shows that over the long run, portfolios with balanced exposure to term and credit spreads outperform portfolios with a concentrated credit spread exposure. The outperformance is most significant during periods of economic weakness and/or financial stress when risks make their "surprising" return to the credit markets.

8.3 RISK PARITY AS A GM HEDGE FUND*

At first glance, the title of this installment of investment insight is plainly confusing. On the one hand, risk parity portfolios are long-only, beta-driven portfolios that balance risk across a diverse array of asset classes to efficiently capture their risk premiums. On the other hand, GM hedge funds are thought of as top-down alpha-generating strategies that capitalize on investment opportunities that result from changing macroeconomic conditions across global financial markets through long and short positions. Their names and descriptions surely imply distinct investment strategies, right?

Confusion regarding the nature of risk parity portfolios would not come as a surprise to us. After all, institutional investors who either have invested in or are considering an investment in, risk parity strategies have generally faced the same question. To what category does risk parity belong: traditional GTAA or alternatives?

The reality is that this is a fair question. For those investors who have not yet considered the similarity between risk parity portfolios and GM hedge funds, now maybe the time to compare risk parity to alternative investments, particularly GM alpha strategies. Just as one cannot judge a book by its cover, one should not judge an investment strategy by its name. In this research note, I will show that indeed these two seemingly different strategies have a lot in common, in terms of their investment objectives, portfolio compositions, and most significantly, their systematic beta exposures. While these commonalities are not generally recognized by most investors, they will probably not surprise those research professionals who have explored hedge fund replication strategies. Nonetheless, the findings detailed in this research note lead to the following question: why replicate hedge fund exposures that

* Originally written by the author in October 2012.

are not thoughtfully constructed when intuitively built, optimal beta portfolios like risk parity are available? Before we present a head-to-head comparison, let us first consider risk parity and GM hedge fund from alternative and opposing perspectives.

8.3.1 Risk Parity as a Hedge Fund Strategy

First, the objective of risk parity is identical to that of hedge fund strategies. They are to deliver stable returns over time, regardless of the broad market condition. Many hedge fund strategies, such as equity market neutral, convertible arbitrage, and merger and acquisition risk arbitrage, achieve this objective by taking long and short positions in positively correlated securities—thus the name, "hedge fund." In comparison, risk parity achieves stable returns in different macroeconomic environments and capital market cycles by balancing the portfolio's risk allocation to asset classes with different sensitivities to growth risks (equities vs. nominal bonds) and inflation risks (real assets vs. nominal assets). In fact, when I developed the concept of risk parity in 2004, my motivation was to identify a beta-driven portfolio with an absolute return objective that can be used as a stand-alone strategy as well as a complement to hedge funds. The fact that many investors measure risk parity strategies relative to commonly used absolute return benchmarks, such as cash plus or CPI plus, is a testament to the "hedge"-like properties of a risk parity portfolio.

A risk parity portfolio generates stable returns over time and under a variety of market conditions through risk diversification. An approach based on risk parity can be viewed as a hedge fund from both an economic and a portfolio perspective. When portfolio risk is properly balanced across a broad array of truly diversified return premiums, the risk parity portfolio is likely hedged against adverse impacts of various macroeconomic shocks.

To illustrate this point, consider a portfolio with 60% of its capital allocated to stocks and 40% allocated to bonds ("60/40 portfolio"). Such a portfolio is heavily exposed to recessionary risk because of its concentrated risk exposure to equities. As a result, it is definitely not structured to meet the objectives of a hedge fund strategy. In contrast, risk parity portfolios target balanced risk contribution from stocks and bonds. Therefore, a significant exposure to high-quality bonds provides a hedge against weakening macroeconomic conditions, while a significant exposure to equities provides a hedge against strengthening macroeconomic conditions. By the same token, a balanced risk allocation between real assets and nominal assets within risk parity portfolios provides a hedge against inflation and deflation risks.

From the perspective of portfolio implementation, a risk parity portfolio is also similar to a long-short hedge fund. This notion may sound strange, but careful consideration of the role correlations play within the portfolio makes it clear. From a technical standpoint, long and short positions serve to hedge a portfolio of positively correlated assets. Conversely, a long-only risk parity portfolio is likely hedged by balancing risk exposures across lowly and negatively correlated assets. This hedging property of risk parity has been abundantly clear over the last 3 years as global financial markets experienced the phenomenon of risk on/risk off and consequently, the correlation between high-quality bonds and risky assets, including equities and commodities, became highly negative. Properly constructed risk

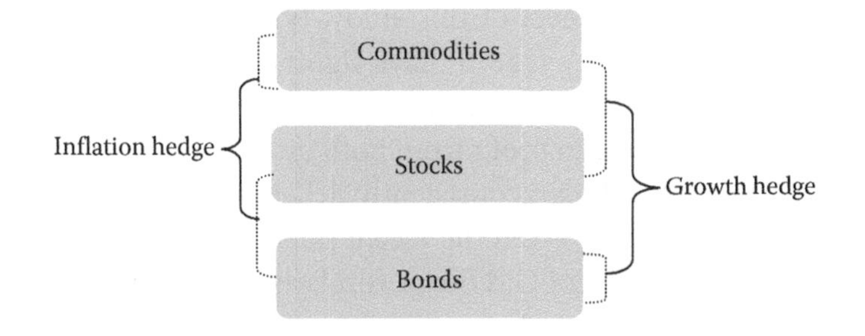

FIGURE 8.13 Hedging against macroeconomic shocks with different asset classes.

parity portfolios, with balanced risk between "safe" and risky assets, behaved like hedge funds in this environment,* with reduced risk and stable returns.

From another perspective, one could argue that risk parity achieves its hedging property more efficiently than traditional hedge funds because it is able to do so in a long-only construct by capturing various forms of market risk premium. Because the risk premiums are positive over the long run, risk parity actually incurs a positive hedging cost. In other words, it appears that one is paid for "having insurance." In contrast, traditional hedge funds that utilize shorting typically have explicit hedging costs.

This hedging property of risk parity is also dynamic when the correlation structure among asset classes evolves with changing macroeconomic conditions. For example, while financial markets are prone to risk on/risk off during periods of low growth and low inflation, their behavior may be markedly different in periods of high and rising inflation, such as those seen in the 1970s. In that environment, correlations between commodities and nominal assets turned negative due to different ways in which inflation shocks impacted real and nominal assets. This phenomenon can probably be coined "inflation-on/inflation-off." A risk parity portfolio with a balanced risk allocation to real and nominal assets can take advantage of this negative correlation, allowing it to be hedged against inflation shocks.

Figure 8.13 shows the schematic linkages among the three asset classes and their respective roles as growth and inflation hedges.

8.3.2 Beta Exposures of Global MacroHedge Funds

Now that we have considered risk parity from a hedge fund perspective, it is time to examine hedge funds, in particular GM hedge funds, from a beta perspective. It is often said that alpha is a zero-sum game. In my view, there are at least two corrections to be made to this statement. First, after management fees, and especially hedge fund incentive fees, alpha is likely to be a negative-sum game. The second correction, which will become evident later in this section, is that while alpha may be negative, the aggregate return of hedge funds can be quite positive if they collectively have positive exposure to beta. In other words, "alpha" might be the result of positive beta.

* See Section 4.3 in Chapter 4 where we discuss the phenomenon of risk on risk off and its implication to asset allocation portfolios.

Since GM hedge funds, the subject of this study, typically use a top-down approach to allocate across asset classes, they tend to have significant exposure to traditional beta relative to other types of hedge funds. As a prior, the beta exposures need not be positive, because hedge funds are expected to profit from both up and down markets. There is little doubt that some GM hedge funds can consistently deliver alpha without a persistent beta bias in their positioning. However, given the recent interest in hedge fund replication, we are interested in the beta exposure of GM hedge funds as a whole. Is there a bias then?

Table 8.11 highlights the excess return, volatility, and Sharpe ratio of the HFR GM Hedge Fund Index from June 1994 to August 2012. The risk-free rate used to calculate excess return is the 3-month T-bill, which realized an annualized return of 3.17% over this period. The HFR GM Hedge Fund Index delivered an annualized excess return of 5.54% with a volatility of 6.33% and a Sharpe ratio of 0.88, which is quite impressive for an index of any strategy.

In addition to the GM Hedge Fund Index, Table 8.11 also shows the return/risk statistics of three market indices for bonds (WGBI, USD-hedged), stocks (MSCI world, USD-hedged), and commodities (DJUBS), as well as two return premiums including smaller size (SML, Russell 2000 less Russell 1000), and higher-beta emerging markets (EMW, MSCI EM less MSCI world). The WGBI index delivered the highest return premium and the lowest risk over this period with a Sharpe ratio of 1.08. The equity and commodity indices as well as the two return spreads all had volatilities between 12% and 16% with low excess returns. As a result, their Sharpe ratios were either flat or only slightly positive.

GM hedge funds do not limit investments in these asset classes. Many are likely to use other investments including currencies, credit, sectors, and other assets. For a simple illustration, we have chosen the three most common betas from the public markets in addition to the two return premiums of size and higher-beta equity markets for a few reasons. First, these betas are commonly used, liquid, and more or less independent; a simple analysis can be transparent and revealing. Second, even though GM strategies invest in other asset classes, they may nevertheless have some exposure to these return premiums. For example, many funds may have exposure to carry trades in currencies, which has positive exposure to equity beta.

A simple way to shed some light on the GM Index's beta exposure is to review the correlation of its returns with the return premiums listed in Table 8.11. Table 8.12 shows this correlation matrix. It is noted that the GM Index is positively correlated with all five betas. In addition, the exhibit shows that the bond index was negatively correlated with the other four betas while the correlation among the four, all of which are more or less linked to economic growth risk, are all positive.

TABLE 8.11 Annualized Excess Return, Volatility, and Sharpe Ratio of HFR GM Hedge Fund Index and Risk Premiums of Traditional Asset Classes

	GM	WGBI	MSCI	DJUBS	SML	EMW
Excess return	5.54%	3.16%	0.86%	2.38%	−0.16%	2.72%
Volatility	6.33%	2.94%	14.74%	16.16%	11.94%	15.69%
Sharpe ratio	0.88	1.08	0.06	0.15	(0.01)	0.17

TABLE 8.12 Correlation Matrix of Returns

	GM	WGBI	MSCI	DJUBS	SML	EMW
GM	1.00	0.14	0.36	0.37	0.24	0.39
WGBI	0.14	1.00	−0.23	−0.13	−0.19	−0.13
MSCI	0.36	−0.23	1.00	0.34	0.13	0.32
DJUBS	0.37	−0.13	0.34	1.00	0.16	0.42
SML	0.24	−0.19	0.13	0.16	1.00	0.31
EMW	0.39	−0.13	0.32	0.42	0.31	1.00

While correlations are illuminating, a regression analysis illustrates the true significance of the beta exposures. Specifically, we regress the monthly excess returns of the GM Index against the monthly excess returns of the five betas. This is akin to style analysis of mutual fund performance based on index returns.* In this research note, I focus our analysis on the entire sample period to study the overall exposures of the index. In future research we will expand the analysis to subperiods and dynamic exposures.

The first regression uses just three market betas, covering bonds, stocks, and commodities. Table 8.13 shows the regression coefficients, their t-stats, and the adjusted R-squared of the regression. All beta exposures are significant, with t-stats above 4. The alpha, about 26 bps per month, is also significant but its t-stat is only slightly above 2. The R-squared is around 25%, so 75% of the GM Index's return variation, or alpha, cannot be explained by the beta indices.

Based on the regression, we can now construct a GM replication beta portfolio with the coefficients as the portfolio weights, thus separating beta and alpha under this framework. Table 8.14 shows the results of the two components. The beta portfolio, labeled $\beta 3$, returned 2.43%, with risk at 3.25% and a Sharpe ratio of 0.75. The alpha portfolio's risk, labeled as $\alpha 3$, is much higher at 5.43%, however its return is only 3.05% and as a result, the Sharpe ratio is 0.56. By comparing the two, one concludes that the alpha portion accounts for much of the risk of the GM returns, but with lower risk-adjusted returns.

The second regression utilizes all five beta factors. Table 8.15 shows the regression coefficients, their t-stats, and adjusted R-squared. Again, all beta exposures are significant. The alpha, 23 bps per month basis, is barely significant. The R-squared increases from 25% to

TABLE 8.13 Regression Results with Three Beta Indices

	Coefficient	t-Stat
Alpha	0.26%	2.33
WGBI	0.55	4.24
MSCI	0.14	5.02
DJUBS	0.12	4.81
Adjusted R^2	25.4%	

* Unlike style analysis for long-only mutual funds, in which its weights have to be positive and sum to one, our regression is unconstrained since in theory hedge funds can be short and use leverage.

TABLE 8.14 The Beta and Alpha Components of GM with Three Market Indices

	GM	β3	α3
Excess return	5.54%	2.43%	3.05%
Volatility	6.33%	3.25%	5.43%
Sharpe ratio	0.88	0.75	0.56

32%. This shows that GM funds, in aggregate, have not only consistent positive exposure to the broad stock/bond/commodity markets, but also consistent positive exposure to high-equity betas in both small cap and emerging markets stocks.

In addition to these factors, other exposures that would likely explain the GM returns are systematic factors such as currency carry, equity value, yield curve slope, etc. An extensive analysis including these additional factors will likely further increase the R-squared of the regression.

Based on this regression, we decompose the GM returns into another set of beta and alpha portfolios. Table 8.16 shows their return, risk, and Sharpe ratios. The β5 portfolio represents a slight improvement over the β3 portfolio with a return of 2.86%, risk of 3.68%, and a Sharpe ratio of 0.78. The α5 portfolio is a little worse than the α3 portfolio. Its risk is 5.15%; the return is only 2.61% and the Sharpe ratio declines from 0.56 to 0.51.

8.3.3 GM Hedge Fund Replication and Risk Parity Portfolios

Our analysis reveals that significant beta exposures exist in the GM Hedge Fund Index. This explains why one of the alternatives to alternatives is the hedge replication strategy, capturing systematic or collective exposures embedded in hedge funds with traditional assets. Based on our analysis, the β3 and β5 portfolios can be used as replication portfolios. But are there better alternatives?

There are at least two better alternatives to the simple hedge fund replication. One is to lever the two beta portfolios. The other more efficient way to capture beta premium is to invest in risk parity portfolios.

The idea of levering the beta exposures embedded in the GM hedge funds is straightforward. In our case, both beta portfolios have high Sharpe ratios, but the risks are quite low. To deliver better returns with similar risk to the GM index, we apply leverage to the underlying beta portfolios. Table 8.17 shows the returns of the leveraged beta replication portfolios, with the risks now being identical to that of the GM index. The levered β3

TABLE 8.15 Regression Results with Five Beta Factors

	Coefficient	t-Stat
Alpha	0.23%	2.10
WGBI	0.63	5.01
MSCI	0.12	4.38
DJUBS	0.08	3.29
SML	0.09	2.73
EMP	0.08	3.20
Adjusted R^2	32%	

TABLE 8.16 The Beta and Alpha Components of GM with Three Market Indices

	GM	β5	α5
Excess return	5.54%	2.86%	2.61%
Volatility	6.33%	3.68%	5.15%
Sharpe ratio	0.88	0.78	0.51

portfolio has a return of 4.73% while the levered β5 portfolio has return of 4.93%; both are just slightly below the index return of 5.54%. Table 8.17 also lists the portfolio weights for the two beta portfolios. Their total leverage is between 150% and 200%.

While hedge fund replication approaches provide investors with systematic exposure to betas, it is not obvious that the weighting of these exposures is optimally efficient. The beta exposures of the GM index are the aggregate exposures of many individual hedge funds. It would require a leap of faith to think that the collective actions of hundreds of managers should lead to consistently more efficient beta exposure. If it does not, then why not use a risk parity approach to build an efficient beta portfolio?

Figure 8.14a and b shows the risk allocations of two replication beta portfolios. In portfolio β3, stocks and commodities dominate the risk allocation while the bond allocation is only 14%. The picture is the same in portfolio β5. The bond risk allocation is only 12% while the reduction in stocks and commodities was replaced by risk allocation in small cap and emerging markets stock beta factors. It is evident that the beta exposures in the GM index are skewed heavily to risky assets or equity-like risks. While GM hedge funds may be much more risk-balanced than traditional asset allocation portfolios, on average their beta exposures fall short of being "hedged" against adverse macroeconomic shocks, leaving room for improvement.

To illustrate the difference, we build two risk parity portfolios, one based on the three beta factors, and the other on the five beta factors* corresponding to the two sets of beta exposures in the regression analysis. The results are shown in the last two columns of Table 8.17. Again, both risk parity portfolios target the same risk level as the GM index. The two

TABLE 8.17 Leveraged Beta Replication Portfolios and Risk Parity Portfolios

	GM	Levered β3	Levered β5	RP3	RP5
Excess return	5.54%	4.73%	4.93%	5.88%	5.39%
Volatility	6.33%	6.33%	6.33%	6.33%	6.33%
Sharpe ratio	0.88	0.75	0.78	0.93	0.85
WGBI		107%	108%	148%	127%
MSCI		27%	20%	24%	17%
DJUBS		23%	14%	21%	13%
SML			15%		21%
EMW			14%		13%

* We target the same risk contribution from the underlying three, or five betas, respectively. The solutions incorporate the sample period volatilities as well as correlations. Similar to the two replication beta portfolios, these risk parity portfolios are based on in-sample risk inputs.

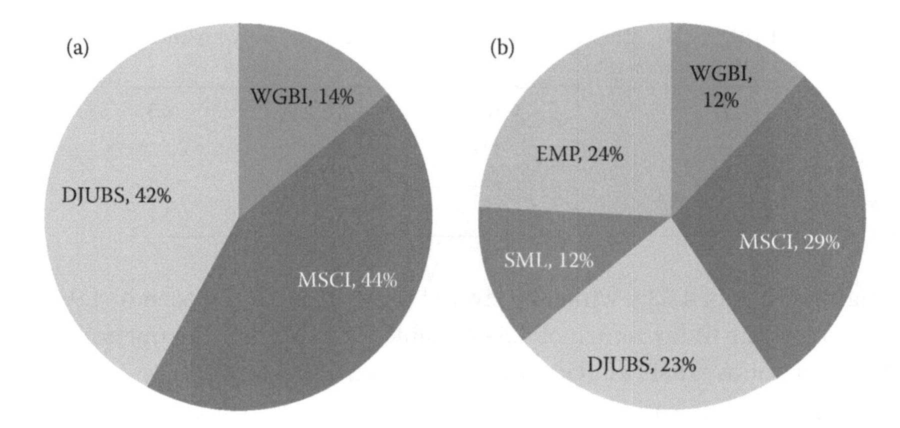

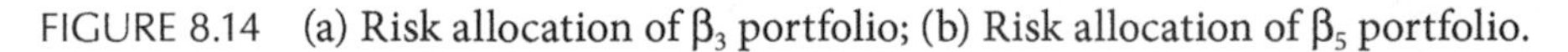
FIGURE 8.14 (a) Risk allocation of β_3 portfolio; (b) Risk allocation of β_5 portfolio.

returns are superior to those of the levered replication beta portfolios and they are now close to that of the GM index. The leverage ratios of both risk parity portfolios are close to 200%. Among, the four beta portfolios and the GM index, the best-performing portfolio turns out to be the most simple of all RP3—the risk parity portfolio with balanced beta exposure in stocks, bonds, and commodities.

8.3.4 Conclusion

In 2006, I presented risk parity at a conference in Scottsdale, Arizona, in which Professor Andrew Lo was also a speaker. His topic was hedge fund replication (Lo, 2006). During the conference, we discussed how both approaches can provide alternative beta portfolios to investors who are not satisfied with traditional asset allocation approaches.

It is evident now that even among these two alternatives, there are commonalities and differences, which we have discussed in this note. Here is what we have learned about the two approaches and their relationship to GM hedge funds. First, risk parity portfolios and GM hedge funds have a lot in common. Risk parity can be viewed as a GM hedge fund from the perspective of having similar investment objectives as well as being hedged from shocks in the macroeconomic condition.

Second, GM hedge funds as a whole have significant beta exposure to traditional asset classes. Moreover, based on data from 1994 to 2012, the beta portion of the GM index has outperformed the "alpha" portion of the index on a risk-adjusted basis. Third, levered beta replication and risk parity portfolios delivered similar, if not potentially higher returns than the GM index with the same risk.

Fourth, it is our experience that adding systematic factor exposures, such as currency carry, equity value, and yield curve slope, to a risk parity portfolio would gain further advantage over the GM index. Fifth, we emphasize that risk parity has a clear investment goal with transparent portfolio implementation. Hedge fund index replication, on the other hand, is subject to many additional uncertainties outside of investment risks. Among them are hedge fund managers' collective biases and estimation errors in the replication process.

To conclude, the name of a particular investment is not particularly important. Some names can even be misleading. What are relevant are the sources of investment returns and risks and how portfolios are constructed to both capture those investment returns and manage the risks. By this measure, risk parity and GM hedge funds belong together.

8.4 PENSION LIABILITIES AND RISK PARITY*

For quite some time, pension funds, both public and corporate, have not been a happy place to be. However, to use a literary quote: "Happy families are all alike; every unhappy family is unhappy in its own way."

8.4.1 The Unhappiness in Pension Families

The biggest difference between the US state public pensions and the US corporate pensions is probably the way they determine the present value of their future liabilities. For public pensions, the discount rate for future liabilities is based on the expected return of the pension assets. Typically, this is assumed to be between 7% and 8%. The assumption is that over time, the assets could earn those rates of return by investing in strategies that often rely heavily on equity risk premium. The argument for this approach is weak at best even though it is compliant with government accounting standards. The drawback of this approach is now in plain sight as the equity market has underperformed for the last decade while pension liabilities remain largely unchanged. As a result, most public pensions are now severely underfunded (Novy-Marx and Rauh, 2009).

In contrast, the Pension Protection Act of 2006 requires corporate pension plan sponsors to evaluate pension liabilities more accurately with high-quality corporate bond yields rather than the expected return on plan assets. This lower discount rate leads to a higher and more realistic estimate of the present value of liabilities, thus forcing corporations to either increase pension contributions or seek liability-matching investment strategies, or both. These efforts have helped corporate pensions to better cope with the perfect storm of declining interest rates and low equity returns. Although most corporate pension plans are also underfunded, corporate pensions are in better shape than their public counterparts.

Going forward, reducing the funding gap remains a daunting challenge for plan sponsors. With the current environment of low interest rates—a result of the global financial crisis and the subsequent subpar economic recovery, public pensions could not possibly afford to bring the discount rate closer to reality while corporate pensions are also reluctant to fully embrace liability-matching investment strategies. To many, a continuing booming equity market, absent the unpalatable options of increasing contributions or reducing benefits, seems the only chance to get out of the current predicament. Or so it seems. But what if equity and equity-like assets underperform again as the global economic recovery falters?

In order to find appropriate solutions, it helps to find the sources of the problems. In this research note, I highlight the different challenges facing public and corporate pensions and propose different solutions to match pension liabilities. In both cases, a risk parity approach can be an effective investment strategy in either an asset-only or an asset-liability management framework.

* Originally written by the author in April 2012.

8.4.2 Public Pensions: Finding That 7% Portfolio with Minimum Risk

Even though the actuarial practice of using projected rates of return on assets instead of market-based interest rates as the discount rate for future liabilities is questionable, it is hard to see plan sponsors changing the practice any time soon. So within the current sub-optimal framework, the challenge is finding investment strategies that achieve the expected return target for the least amount of risk. In essence, the liability is now being "modeled" as a risk-free investment with the targeted return. Given the fact that no such investment exists and thus it cannot be hedged, the next best thing is finding investments that meet the same return (possibly a little higher to ensure an higher probability of meeting the objective) with minimum volatility, i.e., investments with the highest Sharpe ratio (risk-adjusted excess return).

Risk parity strategies that balance the risk contributions of different assets and/or different uncorrelated return premiums represent a suitable solution. Due to proper risk diversification, the long-term Sharpe ratio of risk parity portfolios is expected to be higher than that of traditional asset allocation portfolios. In addition, since risk parity portfolios are less sensitive to different macroeconomic environments, its returns are more robust over market cycles. Volatile return patterns embedded in portfolios concentrated in equity risk inevitably lead to the funding status of the plan to be volatile over time. Large swings in the plan's funding status expose plan sponsors to political and legislative risks, as well as investment risks. Risk parity can help mitigate these risks by providing more stable returns under different market environments.

I shall use a simple stock/bond example to illustrate how risk parity portfolios can reduce these risks. Given the level of current interest rates and volatilities of risky assets, I assume a return of 2% and a risk of 5% for high-quality bonds and a return of 8% and a risk of 20% for stocks. I also assume a correlation of 0.1 between the two asset classes. Those assumptions would imply an equal Sharpe ratio of 0.4 for both asset classes, which is not inconsistent with their long-term averages. Given these low expected returns, partly due to the current risk-free rate in the United States, one can only realistically target a rate of return that is lower than the 7%–8% range with just these two asset classes without extremely high-equity allocation or high leverage. I set it to be 6.5%.

As shown in Table 8.18, to achieve the 6.5% return target, the traditional asset allocation portfolio would have to hold 75% of the plan assets in stocks and the remaining 25% in bonds. The risk of this 75/25 portfolio is 15.2%, resulting in a Sharpe ratio of 0.43. On the

TABLE 8.18 Two Asset Allocation Portfolios with the Same Expected Return at 6.5%

	Traditional	Risk Parity
Stocks	75%	41%
Bonds	25%	164%
Expected return	6.5%	6.5%
Volatility	15.2%	12.1%
Sharpe ratio	0.43	0.54

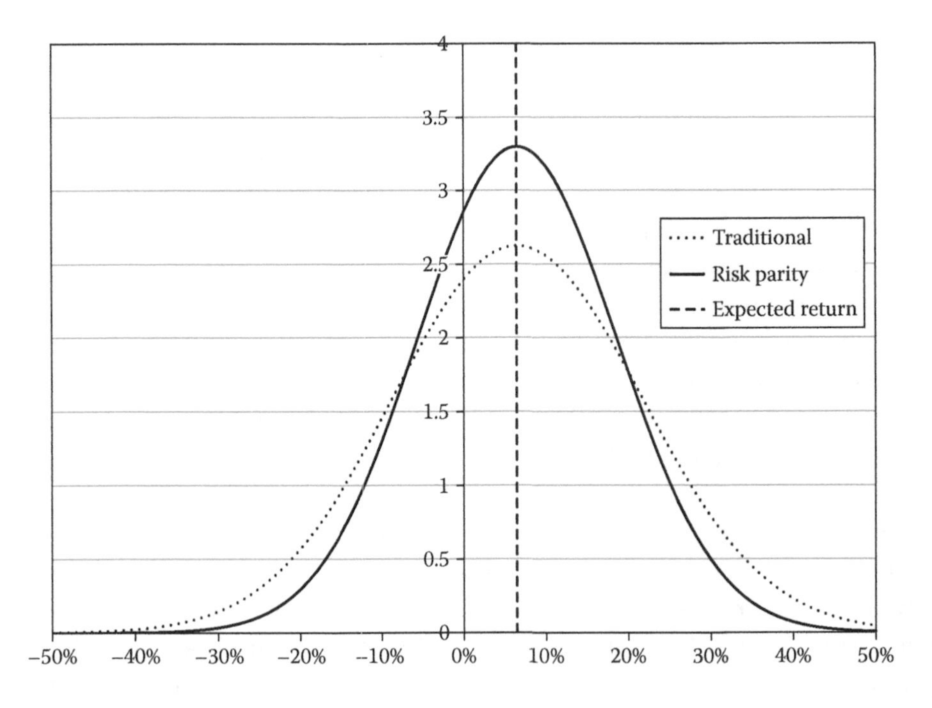

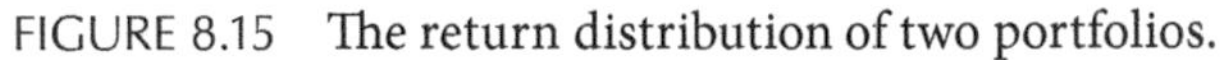
FIGURE 8.15 The return distribution of two portfolios.

other hand, the risk parity portfolio can achieve the same 6.5% return by allocating 41% to stocks and 164% to bonds (4:1 ratio of bonds vs. stocks). Even though it is leveraged at 205%, the portfolio risk is lower, at 12.1% and the Sharpe ratio increases to 0.54.

As shown in Figure 8.15, the lower volatility of risk parity results in a narrower return distribution around the mean and over time. This suggests its deviation from the 6.5% target will be small than that of the traditional asset allocation portfolio. Thus, risk parity can help public pension plans achieve the targeted returns with lower risk, and a more consistent plan funding status.

8.4.3 Corporate Pensions: Using Risk Parity in an Asset-Liability Framework

Since the discount rate applied to corporate pension liabilities changes with the averages of certain high-grade corporate bond yields, corporate pension liabilities are to a large extent marked to market. These changes in the present value of the liabilities would flow through to the corporate income statement, and as a result, they are creating demand for liability-driven investment (LDI) strategies to help manage income stability. A cursory look would indicate that risk parity is also a better fit for LDI strategies than traditional asset allocation approaches, since both have significant notional exposure to interest rate duration.

While this statement is certainly true, it represents an incomplete picture. For corporate pensions, the objective is to minimize the surplus (asset minus liability) risk. If we assign a portion of the fixed-income exposure in the risk parity portfolio to match the liabilities, then the residual portfolio would no longer be risk parity in the surplus risk space.

Within the framework of asset liability management, the optimal investment strategy on the asset side consists of a liability-matching portfolio and a risk asset portfolio (Waring and Duane, 2009). The liability-matching portfolio is dictated by company-specific, or for that matter, agency-specific liabilities.* On the other hand, the risk asset portfolio is quite general, and it is designed to generate surplus return relative to the liabilities with low surplus risk. These goals are consistent with risk parity strategies and hence we should use risk parity for the risk asset portfolio in corporate pension plans. Therefore, the overall assets of corporate pensions can be invested in some combination of liability-matching portfolios and risk parity portfolios as well as other return-generating strategies.

To illustrate the point, I use a corporate bond portfolio as the liability benchmark for corporate pension funds and combine it with the risk parity portfolios discussed above. For the moment, I assume the plan is fully funded at 100%. Because a risk parity portfolio is scalable in terms of its risk/return and leverage, we denote the unlevered risk parity portfolio with 20% in stocks and 80% in fixed income as RP1. Then RP1/2 would be 10% in stocks and 40% in fixed income.

Table 8.19 presents three cases of the total portfolio including both the liability-matching portfolio in 100% corporate bonds with a risk parity portfolio. The case A with RP0 is the fully matched portfolio with 0% investment in risk parity hence zero surplus risk. The case B has RP1/2—50% of the unlevered risk parity and the case C has RP1—the unlevered risk parity. The expected surplus return and risk are from the risk parity portfolios and they are scaled by the percentage of investment in risk parity.

I shall make several remarks about these recommendations. First, I have assumed that a perfect liability-matching portfolio can be found. While this might not be the case in reality, one should nevertheless strive to find a matching investment portfolio that minimizes the surplus risk to the liabilities. Ideally, the surplus risk and return of the liability-matching portfolio should be small compared to the return-generating risk asset portfolio.

Second, for a fully funded plan, the choice A with RP0—just the liability-matching portfolio, has zero surplus risk and it might be the optimal one. However, if a plan sponsor is ever in such an envious position and is willing to take on some surplus risk, I think spending that risk budget on risk parity as proposed in case B and C in Table 8.19 is a better way than splurging on either stocks or bonds.

TABLE 8.19 Example of Corporate Pension Portfolios Using Liability-Matching Asset Portfolio and Risk Parity

	A	B	C
	Portfolio with RP0 (%)	**Portfolio with RP1/2 (%)**	**Portfolio with RP1 (%)**
Corporate bonds	100	100	100
Stocks	0	10	20
Bonds	0	40	80
Surplus return	0	1.6	3.2
Surplus risk	0	3	5.9
Leverage	100	150	200

* These could be pension funds, but also insurance companies, or any other entities with future liabilities.

Third, for the recommendation B and C, one may feel that the total portfolio is now tilted toward bonds, violating the risk parity principle. This is of course not the case since one needs to separate the liability-matching bond allocation from the risk parity portfolio, which now follows the risk parity principle in the surplus risk space under the framework of asset-liability management.

Finally, there is a strong argument against allocating too much equity risk to the risk asset portfolio, which seems to be the prevailing practice in the industry. This is because by doing so, the equity risk in the risk assets is then highly correlated with the growth risk in the overall economy that affects the health of corporate sector at large. If economic growth weakens, the fundamentals of most corporations will likely deteriorate and the loss of equity investment in pension funds can make the matter worse for the company as a whole. If one follows this diversification argument to its logical conclusion, it is probably sensible to tilt the risk asset portfolio toward investments that are countercyclical, such as high-quality bonds. Of course, a detailed analysis and recommendation requires a holistic evaluation of a company's business operation and its pension assets combined.

8.4.4 Underfunded Corporate Pensions

Given the fact that most corporate pension plans remain underfunded, it is important to ask how the proposed portfolios in Table 8.19 change in that case. Can we still implement a liability-matching portfolio? Do we necessarily have to take on more equity risk to close the funding gap? Surprisingly, we need not make much change to the portfolios if we adhere to the asset-liability framework. The only change required is to increase the leverage of the liability-matching portfolio.

Let us use a concrete example and suppose the plan is 80% funded. In other words, the asset-liability ratio is 80%, or reciprocally the liability asset ratio is 125%. To minimize the surplus risk down to zero, i.e., to match the asset exposure to the liabilities, the asset portfolio exposure needs to levered to 125% of the underlying assets. Now that the risk is zero, however, the surplus return is surely negative due to the financing cost of leveraging. Assuming the leveraging cost is 1%, the surplus return will be negative 25 basis points if there is no other investment in the plan.

This investment strategy, shown in Table 8.20 as a, has no surplus risk, but it is destined to remain underfunded forever, because the asset return is always lower than that of the liabilities.* In this example, the leverage ratio on assets is 125% in order to compensate for the underfunded status of the plan. This could be a suitable solution if the plan sponsor makes additional contributions to close the funding gap and moves the plan to match liabilities.

As shown in Table 8.20, similar adjustments are made to the other two investment proposals B and C, which combine the levered liability-matching portfolio in this case with the risk parity portfolios of difference scales. The surplus returns are reduced by the financing costs, the surplus risks remain the same, and the leverage ratios of the total portfolios increase by 25%. The additional row shows the implied surplus Sharpe ratio. The low

* The funding ratio could actually decline while the difference grows because both the numerator and the denominator get larger.

TABLE 8.20 Example of Corporate Pension Portfolios with 80% Funding Ratio Using Liability-Matching Asset Portfolio and Risk Parity

	D	E	F	G
	Portfolio with RP0	**Portfolio with RP1/2**	**Portfolio with RP1**	**Traditional 60/40**
Corporate bonds	125%	125%	125%	0%
Cash	–25%	–25%	–25%	0%
Stocks	0%	10%	20%	60%
Bonds	0%	40%	80%	40%
Surplus return	–0.25%	1.35%	2.95%	2.85%
Surplus risk	0%	3%	5.9%	12.3%
Surplus Sharpe	–	0.46	0.50	0.23
Leverage	125%	175%	225%	100%

tracking portfolio has an implied Sharpe ratio of 0.46 while the high tracking portfolio has a Sharpe ratio of 0.50.

Some readers might question the wisdom of levering the underfunded assets to match the liabilities in order to eliminate the surplus risk. I make three comments on this issue. The first is that the current costs of leveraging are actually quite low. The financing cost depends on the prevailing interest rates and credit qualities of borrowers. Ironically, at today's low interest rates, often cited as the reason against liability-matching schemes, the short-term financing costs of matching liabilities are even lower. For example, UST bill rates are close to zero. Of course, short-term interests could rise in the future. But in today's low interest rate environment the financing cost is not a reason against leverage-based liability matching. One may ask if not now, when?

The second point is that some plan sponsors have always been leveraging the plans by issuing pension-obligation bonds and investing the proceeds along with the pension assets. This form of leverage using long-term financing is not that different from leverage using financial instruments with short-term financing. Of course, one can debate the pros and cons of short-term versus long-term financing, but leverage is leverage.

Finally, this approach is akin to the tax arbitrage that Fischer Black first proposed for corporate pension funds as early as 1980.* But it must be emphasized that all these forms of leveraging are prudent only if it is used for the purpose of liability matching. Issuing pension-obligation bonds and then investing the proceeds in risky assets such as equity is an entirely different story with potentially dubious endings.

With explicit assumptions of surplus return and risk, one can estimate how long it would take for such investment strategies to return the plan to fully-funded status. For a plan that is 80% funded, with 1.35% excess return, it would take a long period of 16 to 18 years on average while the excess return of 2.95% would shorten the average time to 7–8 years. The latter case might seem to be the quicker fix, but the uncertainty would be much greater due to higher surplus risks.

* Fischer Black proposed to invest all proceeds from bond issuance in a diversified portfolio of corporate bonds with similar risk characteristics to that of issuing corporation, for liability-matching as well as tax arbitrage (Black, 1980). In other words, zero surplus risk but positive surplus return.

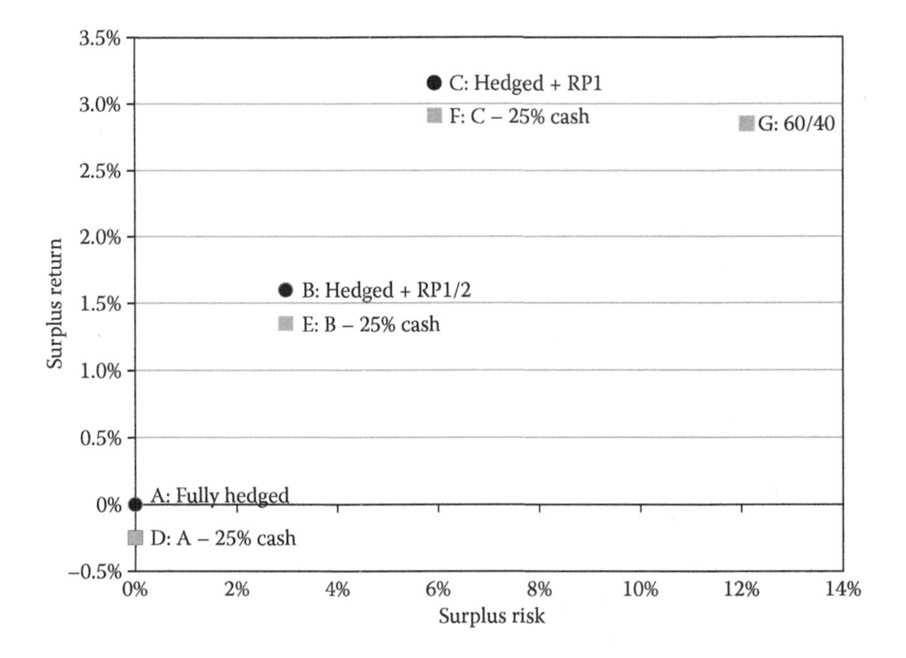

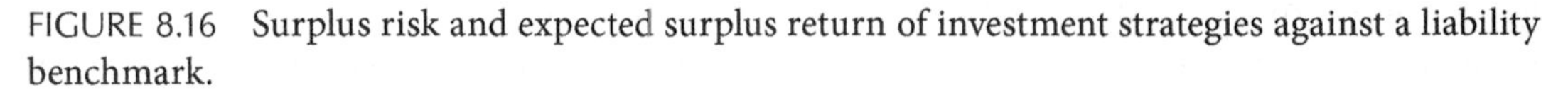
FIGURE 8.16 Surplus risk and expected surplus return of investment strategies against a liability benchmark.

How do these investment strategies compare to a traditional 60/40 portfolio, which is listed as strategy G in Table 8.20? Relative to the liability benchmark of 125% corporate bonds, the 60/40 portfolio is taking an active weight of 60% in equity and of −85% in bonds. For a simple calculation, assume the bonds in the asset portfolio are the same as the bonds in the liability benchmark, then the expected surplus return is 2.85% but the surplus risk is 12.3%. This risk is twice as high as the risk with the investment strategy with RP1 and as a result the surplus Sharpe ratio is only 0.23.

Figure 8.16 plots the surplus risks and returns of seven investments proposed in Tables 8.19 and 8.20. Strategies A, B, and C are for a fully-funded plan and strategies D, E, F, and G are for the unfunded plan. It points to the fact that the traditional 60/40 portfolios are very inefficient as a LDI strategy.

It is extremely rare that active investment strategies would intentionally take on active risks as high as 12.3%. But this is the case with 60/40 portfolios relative to a liability benchmark. The hope of such a strategy is that the risky assets would outrun the liability in the long run. But as we all know too well but sometimes forget, the "hare" does not always win the race against the "tortoise."

8.4.5 De-Risking, but How

Increasingly, many corporate plan sponsors have recognized the significant amount of active risk embedded in a traditional asset allocation portfolio against the liabilities and have taken steps to lower the surplus risk, or to de-risk. But how to de-risk while still preserving positive surplus returns necessary to close the funding gap deserves a careful analysis.

One simple approach is to increase the allocation to fixed income and decrease the allocation to risky assets such as equity in the plan assets. Since the liabilities are all in fixed income, the asset shift to more fixed income will reduce surplus risk. This approach, which I call traditional de-risking, incrementally reduces equity in a 60/40 portfolio and at the extreme makes the asset portfolio 100% fixed income.

The dotted path in Figure 8.17 shows the expected surplus return and risk of this approach for an 80% funded plan. The point G is the 60/40 portfolio seen in Table 8.20, and as we decrease the equity allocation and increase fixed income by the same amount, both risk and return decline. For instance, when the plan asset allocation is 20/80, or 20% in equity and 80% in bonds, the surplus risk is reduced to 4.4% while the surplus return is down to 0.45%. It is apparent that de-risking by traditional approach will have a significant negative impact on the expected surplus return. When the asset portfolio is 100% fixed income, the surplus risk is reduced to 1.25% while the surplus return is negative 75 bps. Since the plan is 80% funded, it is a not a good idea to go 100% fixed income.

An alternative approach, which I call risk parity de-risking, offers better expected surplus returns while reducing surplus risk. The beginning portfolio is still 60/40 but we choose portfolio E—the liability-matching portfolio plus RP1/2 (see Table 8.20) as the target portfolio. As we reduce weight in portfolio G and increase weight in portfolio E, risk parity de-risking lower the surplus risk while maintaining a higher level of surplus return.

The benefit of risk parity de-risking is rather significant. How is this possible? From a Sharpe ratio perspective, for an underfunded plan, traditional de-risking actually reduce the overall Sharpe ratio since a larger fixed-income allocation in the asset mix just "locks in" the negative surplus return relative to the liability benchmark. In contrast, risk parity

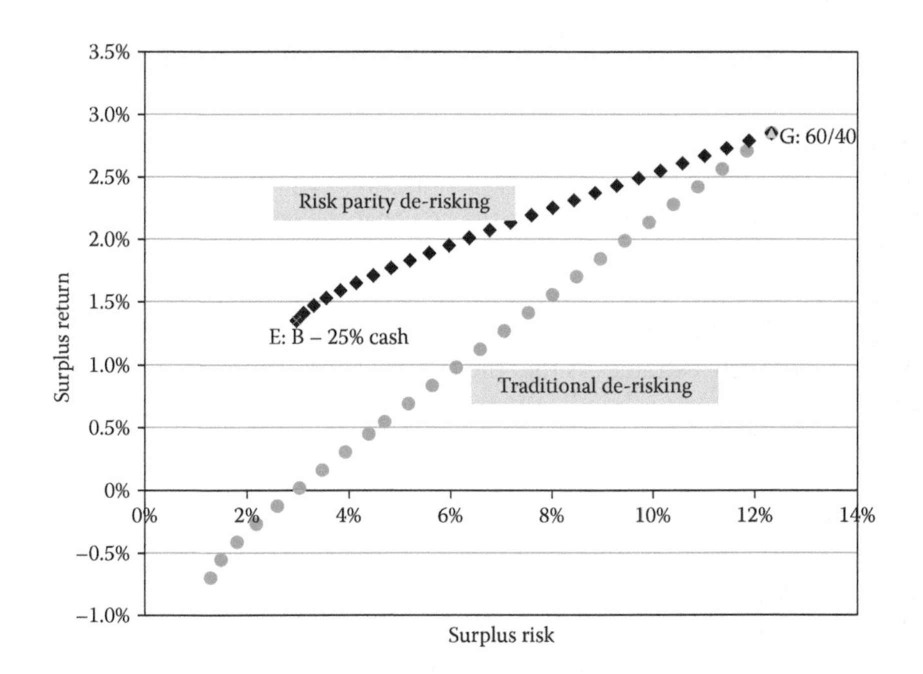

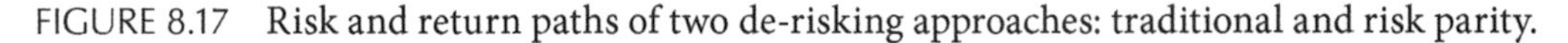
FIGURE 8.17 Risk and return paths of two de-risking approaches: traditional and risk parity.

de-risking increases the Sharpe ratio by both moving closer to liability-matching and increasing the risk-adjusted returns of the risk asset portfolio.

8.4.6 Conclusion

Many pensions are underfunded. It is a challenge to devise investment strategies that could return them to health over time while also protecting against further deterioration. In the current regulatory environment, public pensions and corporate pensions are treated differently and as a result, they may warrant different investment approaches. For the public pension, the approach is investing in strategies that meet the required return with the least amount of risk. For corporate pension plans, the approach is investing in liability-matching portfolios plus risk asset portfolios with an optimal surplus Sharpe ratio.

In this investment insight, I outline the ways that plan sponsors can use risk parity strategies to achieve these investment objectives. For public pension funds, risk parity can provide the targeted return on assets with lower risk, thus reducing the volatility in its funding status. For corporate pension funds, risk parity can be deployed in the risk asset portfolios that are separated from the liability-matching portfolios. In its core, risk parity is an approach to allocating risk efficiently. Whether it is the absolute risk for public pensions or surplus risk for corporate pensions, it provides a way to generate efficient risk-adjusted returns and create the long-term wealth required to meet future pension obligations.

Traditional 60/40 portfolios are ill suited for corporate pension plans because the surplus risk relative to liabilities is extremely high. Traditional de-risking with allocation shifts lowers the surplus risk but incurs significant cost to the surplus return. The alternative risk parity de-risking suggested in this research note can help corporate pensions maintain the level of surplus return while reducing surplus risk. In addition, it moves pension plans closer to a liability-matching framework, which corporations would ultimately adopt.

References

Alonso, N. and E. Qian, Risk parity equity strategy with flexible risk targets, *The Journal of Investing*, Vol. 22, No. 3, 2013, 99–106.

Arnott, R.D., Risk budgeting and portable alpha, *Journal of Investing*, Vol. 11, No. 2, 2002, 15–22.

Barberis, N., Thirty years of prospect theory in economics: A review and assessment, *Journal of Economic Perspective*, Vol. 27, No. 1, 2013, 173–196.

Bernanke, B.S., *The Economic Outlook*, Before the Joint Economic Committee, U.S. Congress, Washington, D.C., May 22, 2013.

Black, F., The tax consequences of long-run pension policy. *Financial Analysts Journal*, Vol. 36, No. 4, 1980, 21–28.

Booth, D.G. and E.F. Fama, Diversification returns and asset contribution, *Financial Analyst Journal*, Vol. 48, No. 3, 1992, 26–32.

Chaves, D., Hsu, J., Li, F. and O. Shakernia, Risk parity portfolio vs. other asset allocation heuristic portfolios, *Journal of Investing*, Vol. 20, No. 1, 2011, 108–118.

Chow, G. and M. Kritzman, Risk budgets, *Journal of Portfolio Management*, Vol. 27, No. 2, 2001, 56–60.

Federal Reserve Board's Semiannual Monetary Policy, Testimony of Chairman Alan Greenspan. Report to the Congress Before the Committee on Banking, Housing, and Urban Affairs, U.S. Senate. February 16, 2005.

Field of Dreams, directed by Phil Alden Robinson, Universal Pictures, USA, 1989.

Gorton, G. and G. Rouwenhorst, Facts and fantasies about commodity futures, *Financial Analyst Journal*, Vol. 62, No. 2, 2006, 47–68.

Grinold, R.C. and R.N. Kahn, *Active Portfolio Management*, McGraw-Hill, New York, 2000.

Hallerbach, W.G., Decomposing portfolio value-at-risk: A general analysis, *Journal of Risk*, Vol. 5, No. 2, 2003, 1–18.

Hallerbach, W.G., Disentangling rebalancing return, *The Journal of Asset Management*, Vol. 15, 2014, 301–316.

Inker, B., The hidden risks of risk parity portfolios, *GMO White Paper*, March 2010.

Inker, B., The dangers of risk parity, *Journal of Investing*, Vol. 20, No. 1, 2011, 90–98.

Jaschke, S.R., The Cornish-Fisher expansion in the context of delta-gamma-normal approximation, *Journal of Risk*, Vol. 4, No. 4, 2002, 33–52.

Jurczenkoa, E., Michelb, T., and J. Teiletchec, Generalized risk-based investing, March 2013, SSRN: http://ssrn.com/abstract=2205979

Kung, E. and L.F. Pohlman, Portable alpha: Philosophy, process, and performance, *Journal of Portfolio Management*, Vol. 30, No. 3, 2004, 78–87.

Lee, I.M., Eight decades of risk parity, *Dimensional Fund Advisors*, August 2011.

Lee, W., Risk based asset allocation: A new answer to an old questions? *Journal of Portfolio Management*, Vol. 37, No. 4, 2011, 11–28.

Litterman, R., Hot spots and hedges, *Journal of Portfolio Management*, December 1996, 52–75.

Lo, A., *Alternative Investment Conference (AIC)*, Scottsdale, Arizona, December 6–8, 2006.

Lohre, H., Neugebauer, U., and C. Zimmer, Diversified risk parity strategies for equity portfolio selection, *Journal of Investing*, Vol. 21, No. 3, 2012, 111–128.

Mackenzie, M., Wigglesworth, R., and Foley, S, *Markets: The Ghosts of '94*, March 19, 2013.

Maillard, S., Roncalli, T., and J. Teiletche, The properties of equally weighted risk contribution portfolios, *Journal of Portfolio Management*, Vol. 36, No. 4, 2010, 60–70.

Mina, J. and A. Ulmer, Delta-gamma four ways, *RiskMetrics Group*, 1999.

Montier, J., No silver bullets in investing (just old snake oil in new bottles), *GMO White Paper*, December 2013.

Novy-Marx, R. and J.D. Rauh, The liabilities and risks of state-sponsored pension plans, *Journal of Economic Perspectives*, Vol. 23, No. 4, 2009, 191–210.

Qian, E., On the financial interpretation of risk contribution: Risk budgets do add up, *Journal of Investment Management*, Vol. 4, No. 4, 2006, 1–11.

Qian, E., Risk parity and diversification, *Journal of Investing*, Vol. 20, No. 1, 2011, 119–127.

Qian, E., Diversification return and leveraged portfolios, *Journal of Portfolio Management*, Vol. 38, No. 4, 2012, 14–25.

Qian, E., Pension liabilities and risk parity, *Journal of Investing*, Vol. 21, No. 3, 2012, 93–101.

Qian, E., Are risk parity managers at risk parity? *Journal of Portfolio Management*, Vol. 40, No. 1, 2013, 20–26.

Qian, E., On the holy grail of "upside participation and downside protection," *Journal of Portfolio Management*, Vol. 41, No. 2, 2015, 11–22.

Qian, E., Bryan, B., and K. Yang, Diversifies factor premiums, *Investment Insight*, PanAgora Asset Management, August 2013.

Roncalli, T., *Introduction to Risk Parity and Budgeting*, Chapman & Hall/CRC Press, Boca Raton, FL, 2014.

Sharpe, W., Determining a fund's effective asset mix, *Investment Management Review*, December 1988, 59–69.

Sharpe, W., Asset allocation: Management style and performance measurement, *Journal of Portfolio Management*, Vol. 18, No. 2, 1992, 7–19.

Sharpe, W., Budgeting and monitoring pension fund risk, *Financial Analyst Journal*, Vol. 58, No. 5, 2002, 74–86.

Wall Street Journal, Open Letter to Ben Bernanke, November 15, 2010. http://blogs.wsj.com/economics/2010/11/15/open-letter-to-ben-bernanke/

Wander, B.H., De Silva, H., and R.G. Clarke, Risk allocation versus asset allocation, *Journal of Portfolio Management*, Vol. 29, No. 1, 2002, 9–30.

Waring, M.B. and W. Duane, An asset-liability version of the capital asset pricing model with a multi-period two-fund theorem, *Journal of Portfolio Management*, Vol. 35, No. 4, 2009, 111–130.

Winkelmann, K., Improving portfolio efficiency, *Journal of Portfolio Management*, Vol. 30, No. 2, 2004, 23–38.

Index

V

Y

Z

An environmentally friendly book printed and bound in England by www.printondemand-worldwide.com

This book is made of chain-of-custody materials; FSC materials for the cover and PEFC materials for the text pages.